What You *Really* Need to Know Before Anatomy, Physiology, and Microbiology

A science primer for pre-nursing & allied health students

By Lesley E. Blankenship-Williams, Ph.D.

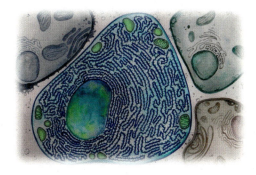

What You Really Need to Know Before Anatomy, Physiology and Microbiology.

ISBN 978-0-692-48192-9

Copyright © 2015 by Lesley E. Blankenship-Williams

All rights reserved. No part of this publication may be reproduced or transmitted in any form or by any means, electronic, mechanical, or otherwise, without prior permission in writing from the author.

Printed and bound by CreateSpace.

My students are wonderful teachers. I learn more from them than they know. Thank you to all my past, present and future students for enriching my life and teaching me about how people learn.

Table of Contents

PREFACE AND INTRODUCTION	5
CHAPTER MAP	7

Chapter 1: BACK TO THE BASICS OF CHEMISTRY — 8
Atomic structure, elements, electron shells

Chapter 2: CHEMICAL BONDS AND STUFF — 15
Covalent bonds, ionic bonds, hydrogen bonds, polarity

Chapter 3: SOLUTIONS — 27
Solubility in water, pH, acids, bases, pH buffer

Chapter 4: SHORT-HAND ORGANIC CHEMISTRY — 40
Different ways to draw organic molecules, common functional groups

Chapter 5: MANAGING HUMAN ANATOMY — 48
Efficient study approaches for human anatomy

Chapter 6: CARBOHYDRATES AND NUCLEIC ACIDS — 63
Monomer, polymer, sugars, polysaccharides, nucleotides, bases, ATP, DNA, RNA

Chapter 7: PROTEINS — 77
Amino acids, polypeptides, folding, residues, functional protein

Chapter 8: CELLS AND HISTOLOGY — 90
Structure-function relationship, eukaryotic cell, organelles, specialization, tissue

Chapter 9: PLASMA MEMBRANE AND LIPIDS — 106
Fats, steroids, phospholipids, cholesterol, lipid bilayer, membrane proteins, fluid-mosaic model

Chapter 10: ENERGY TRANSFORMATIONS — 118
Laws of thermodynamics, chemical reactions, endergonic, exergonic, redox, ATP, potential energy, cellular respiration

Table of Contents

Chapter 11: MOLECULAR TRANSPORT   138
Diffusion, osmosis, concentration, transport across a membrane, channel and carrier proteins, active and passive transport, vesicles, permeability, sodium-potassium pump

Chapter 12: ELEGANT ENZYMES   161
Enzymes, activation energy, reaction rate, inhibitors, activators, metabolic pathways

Appendix

   Web links for videos   179

   Acknowledgements   180

   Full list of image credits   180

Author's Preface & Introduction

Is your intended destination a nursing, physical therapy, pharmacology, or physician's assistant program?

If *yes*, then this book is probably for you.

The author and her husband at Ano Nuevo State Park in California filming elephant seal rookeries with Palomar College TV Center (2012). *Photo courtesy of Palomar College TV Center.*

To gain admittance to one of these programs, you must excel in the rigorous and challenging courses of Human Anatomy, Human Physiology and Microbiology. Passing with a C grade is no longer enough. Yet up to 50% of enrolled students fail or drop out of these classes (Sturges and Maurer, 2013). In fact, the high failure rate is so alarming, a comprehensive study addressed this very issue – *why are so many students floundering in these classes?*[1] Turns out that the number one obstacle to student success is.....**the material is hard — really, really hard!**

I've been teaching Human Anatomy, Human Physiology and Microbiology at community colleges for over ten years. The Sturges and Maurer (2013) study is accurate. The material is hard and there is lots of it. You need good study skills, ample study time, and you absolutely must be prepared to tackle the difficulty of the content. To put yourself in the best possible position to comprehend the complexities of Human Anatomy, Human Physiology and Microbiology, you need an excellent working knowledge of chemistry and biology.

Before you counter with a *"but I already took a biology (or chemistry) class,"* retort, let me be frank. Just because you took a chemistry and/or biology class doesn't mean you currently possess an excellent working knowledge. Case in point – I earned a bachelor's degree in Pure Mathematics in the year 2000. Ask me to solve a calculus problem right now and I couldn't without some serious refreshing.

Therefore, if any of the following criteria apply to you, this book is probably worth your time.

(1) You completed either a college-level chemistry or biology course one year ago or more.

(2) You earned less than 90% on your general biology or chemistry exams.

(3) You struggled with the concepts when you took the biology or chemistry course. You memorized rather than understood the material.

If your science foundation is weak, much of the material covered in Human Anatomy, Human Physiology and/or Microbiology will be over your head. Your only hope is to remediate in basic chemistry and biology quickly or risk drowning in the current material. To expect the instructor re-teach all the prerequisite biology and chemistry is foolishness. It is your responsibility to know the prerequisite biology and chemistry material coming into the class, and there is not enough instructional time for the professor to adequately teach the science you should already know.

---

[1] The study published by Sturges and Maurer (2013) addressed *Human Anatomy* and *Human Physiology* courses, but not *Microbiology*. However, the same conclusions apply to *Microbiology*.

So what is a student to do? This predicament is precisely why I wrote this book.

Until now, there were no widely available resources that quickly and efficiently equipped students with the foundational biology and chemistry required to excel in Human Anatomy, Human Physiology and Microbiology. My goal is to quickly get you up to speed by providing a solid working knowledge of the basic chemistry and biology that you *really* need to understand anatomy, physiology and microbiology. This book is not intended to replace a biology or chemistry class – it is assumed that you have already taken one or both courses – but is meant to be a straightforward, quick, pointed, and engaging refresher.

The topics presented in this book were chosen by faculty and students. Twenty faculty who teach Human Anatomy, Human Physiology or Microbiology at community colleges described the biology and chemistry topics that students *should* know (but often do not) before starting one of these classes. Approximately 300 students enrolled in these courses were surveyed about the same list five weeks into the semester – *what topics did students need to review the most?* From the student surveys and faculty input, twelve topics were selected for inclusion in this book. Not every chapter applies to every class, and the chapter map on the next page will help you determine which chapters are applicable to the course you are currently enrolled in.

This is no ordinary textbook; it is a workbook. The emphasis is on solving problems and answering questions. This format gives you *something to do* (i.e. active learning versus passive learning) and also provides opportunities for self-assessment. After all, attempting to solve a problem is a great way to gage your understanding. An answer key is provided at the end of the chapter and the more difficult problems also include explanations.

The more pages in a book, the more it costs to print. To cut down on pages (and therefore cost), some of the information is presented by video tutorials rather than text. These video tutorials link to QR codes embedded into each chapter. The QR codes can be read by any smartphone or tablet with a **QR Reader app** installed. Before diving into this workbook, make sure your smartphone (or tablet) is properly equipped by installing a **QR Reader** app (it's free!). Once the app is installed, go ahead and open the QR Reader app, and place the camera over the QR code until a video appears. The video will finish my introduction.

If you don't own a smartphone or tablet, the web address for all videos is available on page 179. In lieu of a smartphone or tablet with internet access, you will need access to a computer with an internet connection so you can access the videos while working through this book.

Best of luck in your endeavors.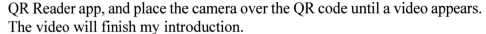

<u>Citation</u>

Sturgers, D. and T. Maurer. (2013) Allied health students' perceptions of class difficulty: The case of undergraduate human anatomy and physiology classes. *The Internet Journal of Allied Health Sciences and Practices*. Volume 11, pp. 1-10.

Which Chapters Should You Complete?

Here is a suggested sequence for the chapters that are most important for the course you are currently enrolled in.

Regardless of which course you are starting or currently enrolled in, work through Chapters 1, 2, 3, and 4 first. These four chapters provide the chemistry foundation necessary to understand all biological principles.

Human Anatomy

After finishing **Chapters 1** through **4**, move onto **Chapters 5** and then **8**.

Human Physiology

After finishing **Chapters 1** through **4**, move onto **Chapters 6**, **7**, **9**, **10**, **11** and **12**.

Chapter 8 may be useful if you would like a refresher on cellular organelles.

Microbiology

After finishing **Chapters 1** through **4**, complete **Chapters 6**, **7**, **9**, **10** and **12**.

The first few pages of Chapter 8 is helpful for an overview of cell types.

Chapter 11 has useful information in reference to permeability and osmosis.

Chapter	Human Anatomy	Human Physiology	Microbiology
1	✓	✓	✓
2	✓	✓	✓
3	✓	✓	✓
4	✓	✓	✓
5	✓		
6		✓	✓
7		✓	✓
8	✓	May be useful	
9		✓	✓
10		✓	✓
11		✓	May be useful
12		✓	✓

Chapter 1: Back to the basics of chemistry

It's elemental, my dear.

Figure 1. A lego dragon is comprised of tens of thousands of legos. A human body is comprised of 7 billion, billion, billion atoms (7 x 10^{27}).

All tangible things are comprised of **atoms**. For example, this book is made of paper. Paper is made from a fibrous molecule called cellulose. Cellulose is a bunch of carbon, hydrogen and oxygen **atoms** linked together in a specific structure. The chair that you sit on, the floor you stand on, the air you breathe, and even your flesh and blood consist of **atoms**. **Atoms** are so very small, you can't see them as individual units even with a classroom microscope. Yet, your body is built from them – much like tiny lego bricks make up fantastically large structures at Legoland (**Figure 1**). Given the fact that humans (well, all things, really) are made up of bazillions of atoms, it's probably pretty important that we get to know them. And not just *kind of* know them. Really know them. Let's get started.

Each atom is comprised of three types of subatomic particles – **protons**, **neutrons** and **electrons**. Each proton has a +1 charge (proton & positive both start with the letter 'p'). Each electron has a -1 charge. Each *neutr*on is *neutr*al (no charge).

Notice that all the protons and neutrons are clustered together in the center in a region known as the subatomic nucleus. The electrons are orbiting the nucleus. In reality, electrons do not orbit the nucleus like planets orbit the sun as shown in **Figure 2**. Rather, electrons exist in a space known as an electron cloud. But in biology, we like to represent electron structures in circular orbits because it helps us predict how those electrons interact with other electron structures, and this is very important for learning about chemical bonds. But more on that later.

Protons and neutrons have a tiny but measurable mass. We call it a dalton (da). The electrons have a weight much less than 1 dalton and so we ignore the weight when calculating the atomic mass of an atom. For example, an atom with 12 protons, 13 neutrons, and 12 electrons would have an atomic mass of

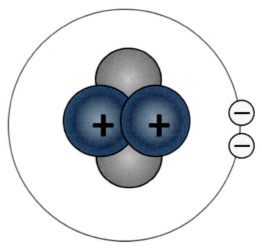

Figure 2. A simplistic representation of a helium atom.

approximately 25 daltons [(12 protons + 13 neutrons) x 1 da = 25 da].

Problem 1

An atom has 5 protons, 5 neutrons, and 5 electrons. In the blank space, re-draw these subatomic particles to correctly represent the atomic structure.

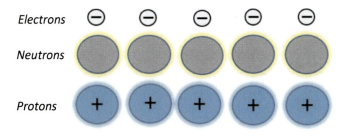

Problem 2

What is the atomic mass of the atom in Problem 1?

Do not confuse the term **element** with an **atom**. If each atom is an individual lego brick, then the *type* of lego brick (e.g. a square red brick) is the element. If you are familiar with building legos, you know that not all lego brick types are used in each structure, and that some bricks are used more than others. In the same way, not all of the naturally occurring elements are used in a structure. Some elements are very common, and some are rare. The four most common elements in biology are hydrogen, oxygen, carbon and nitrogen. In fact, 96% of your body is made up of these four elements!

Problem 3. Lego stores sell legos arranged in bins. One bucket contains purple legos of the same brick model, while another bucket may contain all orange legos of a different brick model.

1. A *bucket* of legos of the same type represents one
   _____.
   a) element
   b) atom

2. A single lego brick represents one _____.
   a) element
   b) atom

6
C
12.01

The periodic table is a wonderfully informative graphic describing each element. Let's use one box from the periodic table as an example. Each element is given a chemical symbol. The chemical symbol for carbon is "C". The chemical symbol may be the first letter(s) of the element's name, but not always! Notice that there are two numbers in this box. **The top number represents the number of protons that defines each element.** Did you read that last sentence? It's an important one.

The number of protons in an atom determines which element that atom belongs to.

The bottom number in the box represents the **atomic mass**. You might notice the bottom number is approximately double the top number. This is because, in most cases, there are equal numbers of protons and neutrons in an atom. Think about a carbon atom. By definition it has to have 6 protons. Therefore it probably has about 6 neutrons. This means the atomic mass is 12 daltons.

On a side note......

Atoms belonging to one element always have the same number of protons, but not necessarily the same number of neutrons. For example, all carbon atoms have 6 protons and most have 6 neutrons. But some carbon atoms have 7 neutrons and a few actually have 8 neutrons. Those variants with a different number of neutrons are called **isotopes**. The atomic mass of a carbon atom with 7 neutrons would be 13. Most carbon atoms weigh 12 daltons, but a few weigh 13 and some even weigh 14 daltons. The atomic mass is the average of all the isotopes and their earthly proportion.

Problem 4

Use a periodic table (find one on the internet) to finish the following table. You will need internet access to look up the element names and symbols.

Element name	Chemical symbol	Number of protons	Atomic mass
Hydrogen	H	1	1.01
	C		
		7	
			16.00
Potassium			
	Na		
		15	30.97
Chlorine			
	Ca		

The number of protons *determines* the element, but it is the number of electrons that determines how an atom can react chemically. How can the periodic table tell you how many electrons are in an atom? It doesn't explicitly tell you. But with this rule of thumb, you can figure it out.

 The number of protons = the number of electrons (at least to start with).

An atom that has 5 protons has 5 electrons. An atom with 19 protons has 19 electrons and so forth.

Electrons occupy a specific region near the subatomic nucleus. We call those locations **shells**. In biology, we represent those shells like Saturn's rings, even though in reality electron behavior is more complex.

Here are the rules:

1. The first shell holds a maximum of 2 electrons.
2. The second shell holds a maximum of 8 electrons.
3. The third shell holds a maximum of 8 electrons.[1]
4. Always fill in the electrons from the ground up. In other words, if you have 11 electrons, 2 would go in the first shell, 8 in the second shell and only 1 in the third shell.

Problem 5

Use the information from Problem 4 to determine how many electrons will go in each shell for the following atoms. As an example, oxygen has been done for you.

O — 3rd shell (maximum 8): [ ]; 2nd shell (maximum 8): 6; 1st shell (maximum 2): 2

C — 3rd shell (maximum 8): [ ]; 2nd shell (maximum 8): [ ]; 1st shell (maximum 2): [ ]

Cl — 3rd shell (maximum 8): [ ]; 2nd shell (maximum 8): [ ]; 1st shell (maximum 2): [ ]

Na — 3rd shell (maximum 8): [ ]; 2nd shell (maximum 8): [ ]; 1st shell (maximum 2): [ ]

---

[1] The third shell can actually hold 18 electrons, but that discussion is beyond the scope of this tutorial. We will stick with 8 electrons as the maximum the third shell can hold.

When we distribute our electrons in a particular shell, we set aside space for them in pairs. Electrons really do need to be in pairs or they behave badly. A quick video will remind you how to distribute the electrons for an atom. Put your smartphone or tablet over this QR code to watch the video. A reminder that you need a QR code reader app installed on your smartphone or tablet.[2]

> *On a side note......*
>
> Ever heard of a *free radical*? **A free radical is the result of an unpaired electron.** Unpaired electrons behave very badly and will stop at nothing to steal a partner electron from somewhere else. Here's the rub - when a free radical steals a single electron, the victimized atom or molecule becomes a free radical itself. And the new free radical sets off to steal a replacement electron. This sets off a chain reaction that can be extraordinarily damaging to our cells as molecules are destroyed in the process. *Antioxidants* are chemicals that can donate that last electron somewhat safely to a free radical, and thus stop the free radical destruction.

One last thing. There is a special name for the outer shell containing electrons - the **valence shell**. A hydrogen atom only has a single electron and therefore uses the first shell only. The first shell for a hydrogen atom is also the valence shell. In contrast, the valence shell for a carbon atom (which has six electrons and therefore uses two shells) is the second shell. Here's our next rule of thumb.

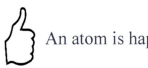 An atom is happy 🙂 (i.e. stable) only when its valence shell is complete!

In Chapter 2 you will learn how atoms respond when their valence shells are incomplete, which is the basis of chemical reactivity and chemical bonds. You're almost done! Try the following problems and then check all of your answers with the answer key on page 14.

Problem 6

Draw the electron structures for each atom. Distribute the electrons correctly into the appropriate shells (as demonstrated in the video).

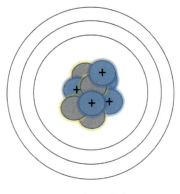

Carbon (C)

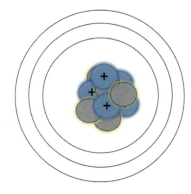

Nitrogen (N)

---

[2] When this video is viewed on a smartphone or tablet, the subtitles may not appear.

Problem 6 Continued

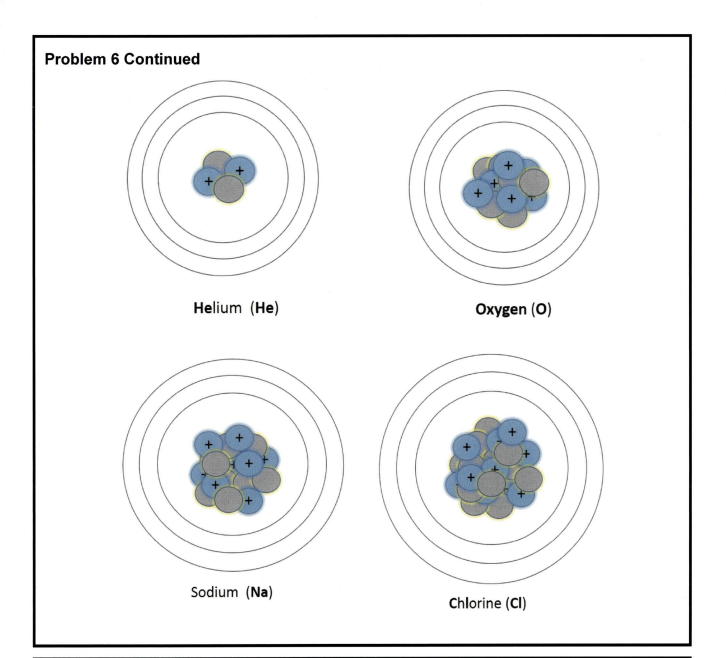

Helium (He)   Oxygen (O)

Sodium (Na)   Chlorine (Cl)

Problem 7

Mark or highlight the valence shell for each atom in Problem 6.

Problem 8

Which atoms in Problem 6 have <u>complete</u> valence shells, if any? Remember, atoms with complete valence shells are stable (or "happy"). 🙂

Chapter 1: Answer key

1.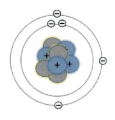

2. 10 daltons

3. 1. a) element   2. b) atom

4.

Element name	Chemical symbol	Number of protons	Atomic mass
Hydrogen	H	1	1.01
Carbon	C	6	12.01
Nitrogen	N	7	14.01
Oxygen	O	8	16.00
Potassium	K	19	39.10
Sodium	Na	11	22.99
Phosphorous	P	15	30.97
Chlorine	Cl	17	35.45
Calcium	Ca	20	40.08

5. Answers given as 1st shell; 2nd shell; 3rd shell

Carbon (C) has 6 electrons – 2; 4; 0

Chlorine (Cl) has 17 electrons – 2; 8; 7

Sodium (Na) has 11 electrons – 2; 8; 1

6 and 7. Valence shells are in **red**.

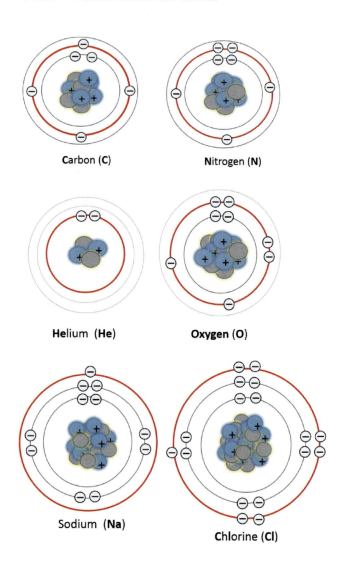

8. Only Helium (He) has a complete valence shell.

Chapter 2: Chemical bonds & stuff

One atom's junk is another atom's treasure.

In Chapter 1, we learned that atoms are only "happy" when their valence shells are complete. Yet, very few elements are naturally equipped with a complete valence shell (helium and neon are two examples of naturally "happy" elements). We don't hear about such stable elements in biology because they do not chemically react.

This chapter is about the elements with incomplete valence shells. Hydrogen, carbon, nitrogen, oxygen, potassium, chlorine, calcium and sodium all have incomplete valence shells. **The number one priority for these atoms is to complete their valence shells!** How? There are two different strategies an atom can use to fill its valence shell.

Strategy 1: Share electrons with another atom. This is a very common solution and the result is a **covalent bond**.

Strategy 2: Donate or accept an electron outright. The result is an **ion**.

A solid understanding of these concepts is crucial to your ability to comprehend basic human physiology and microbiology topics. Let's explore these strategies some more.

The concept of sharing electrons is easily demonstrated with the formation of a water molecule. You will recognize the chemical formula for water as H_2O. H_2O is two hydrogen atoms that share their electrons with a single oxygen atom.

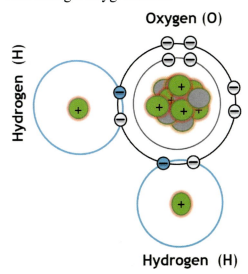

Figure 1. A water molecule. The white electrons belong to the oxygen atom and the blue electrons belong to hydrogen atoms.

If that last sentence was a mouthful, place your smartphone or tablet over this QR code to watch a video of how a water molecule is constructed.

Take a close look at **Figure 1**. Remember that a single covalent bond is actually a shared PAIR of electrons. Usually each atom puts some "skin in the game" by contributing a single electron to the covalent bond. The term *covalent* isn't just mumbo jumbo, but a conjunction of *co–* (as in **co**operate) and *–valent* (which means the **valence** shells). So, a covalent bond is a cooperation between electrons of two valence shells. The pair of electrons will spend some time orbiting one atom, and some time orbiting another atom. The situation is analogous to the parent who buys a single toy to be shared among two siblings. Each sibling takes turns playing with the toy. The toy is the shared pair of electrons and the siblings are the two atoms.

Once a covalent bond is formed between two atoms, we have a **molecule**. A molecule is the structure consisting of two or more atoms held together by a covalent bond. Water is a molecule.

Problem 1

Draw the complete electron structures for NH_3 and CH_4 (in a similar fashion as **Figure 1**). Then, circle each shared *pair* of electrons.

Problem 2

How many covalent bonds are in a H_2O molecule total? _____

How many shared electrons are in a H_2O molecule? _____

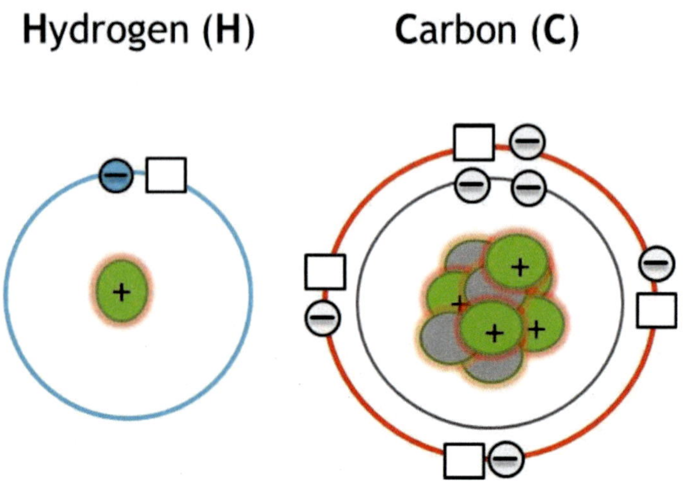

Hydrogen (H)

Carbon (C)

Figure 2. The boxes represent electron vacancies. Filling all vacancies will complete the valence shell.

The *siblings sharing a toy* analogy is a good one when we consider siblings of different ages. Think of a 7 year old and a 3 year old sharing a toy. It is unrealistic to think that the siblings, left to their own devices, will share the toy equally. If the older child wants the toy, then the older child is going to play with the toy more than 50% of the time. The younger child would get the "short end" of the toy.

As we will discuss in a few pages, some elements have a higher affinity for electrons than others (we call this property **electronegativity**). When a covalent bond exists between two atoms of different electronegativities, the electrons are not shared equally.

The number of unpaired electrons in the valence shell is equal to the number of covalent bonds that an atom can form. **Figure 2** shows that hydrogen has a single electron but only needs one more electron to fill its valence shell (the empty box represents the electron vacancy). Thus, hydrogen is capable of forming a single covalent bond. A carbon atom needs four more electrons to fill its valence shell; carbon forms four covalent bonds.

Problem 3

Review the electron structures of oxygen and nitrogen to answer the following questions.

How many covalent bonds will an oxygen atom form? _____

How many covalent bonds will a nitrogen atom form? _____

Drawing in all electrons for all atoms gets cumbersome. Therefore, we just use a single solid line[1] to represent a single covalent bond. **Figure 3** shows the line equivalency for methane (CH_4) and atmospheric oxygen (O_2).

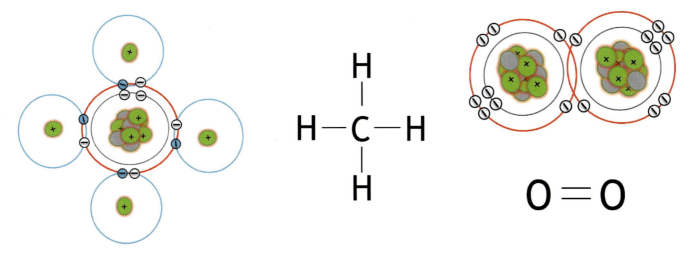

Figure 3. The far left image is the electron structure of CH_4 (methane). Try to identify the four covalent bonds. The middle image is also CH_4 with the covalent bonds represented by single lines. You should be able to superimpose the two images. The top right image is of O_2 (also known as *atmospheric oxygen* to distinguish the molecule from the element oxygen). The bottom right image is the line model of atmospheric oxygen. Note that O_2 has a double covalent bond; there are 4 electrons being shared between the two atoms.

On a side note......

Ever hear of saturated versus unsaturated fatty acids (or fats)? Unsaturated fatty acids are long chains of carbon atoms with at least one double covalent bond between two carbon atoms. A saturated fatty acid is a long chain of carbon and hydrogen atoms with all single covalent bonds (no double covalent bonds). The difference of a double covalent bond (versus a single covalent bond) has dramatic effects on the properties of these fats, as we will see in Chapter 9.

A shared pair of electrons is a *single* covalent bond. But some atoms can share more than two electrons. O_2 is the result of two oxygen atoms sharing four electrons (**Figure 3**). N_2 is when two nitrogen atoms share six (!!!) electrons resulting in a triple covalent bond. Double and triple covalent bonds are stronger (like adding additional ropes to hold two items together), but also fix the atoms in place without the ability to rotate. Use your smartphone or tablet over this QR code for an additional

---

[1] Using a solid line is called the Couper notation. Using dots is called the Lewis dot system (or Lewis notation).

tutorial about multiple covalent bonds and their lack of rotation.

Problem 4

Draw H₂O and NH₃ using the **solid line** model.

Still with me? We've covered the first of two strategies for filling valence shells. Let's explore the second strategy. The second strategy is to give away or accept one or two electrons outright. **In order to exercise this strategy, the element must be <u>REALLY</u> close to having a complete valence shell.** An example will illustrate this.

Figure 4 shows the electron structures of Na (sodium) and Cl (chlorine). Sodium has only one electron in its valence shell. To completely fill the third shell, sodium would need seven (!!) other electrons. That's a tall order (atoms just aren't that giving!). But what if sodium donated its 11th electron? Then only ten would remain and the valence shell (now the second shell) is complete!

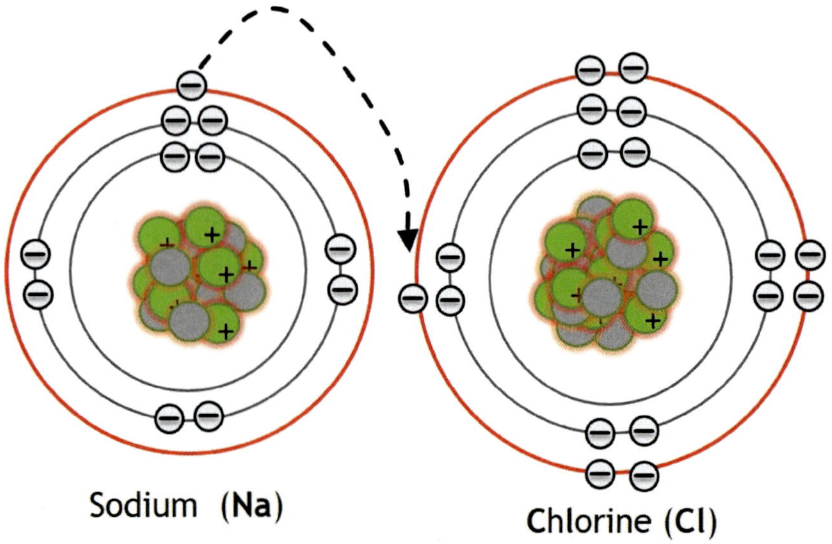

Look at chlorine. Chlorine needs one more electron to complete its valence shell. What are chlorine's options? Chlorine could (and sometimes does) form a single covalent bond to fill its valence shell. But if sodium is offering up a free electron, chlorine is more than willing to take it for its own as shown in **Figure 4**. *Truly, one atom's junk is another atom's treasure!*

When an electron transfer occurs like the one shown in **Figure 4**, each atom no longer has equal numbers of protons and electrons. An atom or molecule with an unequal number of protons and electrons is an **ion**. Ions have either a net positive, or net

Figure 4. Sodium donates its 11th electron to chlorine. This give-take strategy fills both valence shells without any covalent bonds. However, the process creates two ions. Sodium now has a positive charge and is denoted as Na$^+$. Chlorine has a negative charge (and technically adopts a new name called chloride) and is written as Cl$^-$.

negative charge. Place your smartphone or tablet over this QR code to watch the animation of how ions are formed and how their charges are calculated.

Problem 5

Potassium (**K**) has 19 protons. Therefore, potassium is most likely to exist as _____.
(SOLVE this problem using your reasoning skills – don't just look up the answer on the internet).

a) K^- b) K (neutral)   c) K^+ d) K^{2+} e) a molecule with 4 covalent bonds

Problem 6

Calcium (Ca) has 20 protons. Explain why calcium usually exists as a Ca^{2+} ion.

It's important to know which elements are likely to become ions, and which ones are not. Carbon, for example, is not going to be an ion. Neither is nitrogen. Although there is lots of fancy chemistry for predicting which elements become ions, the following rule of thumb will suffice for your anatomy, physiology and microbiology courses.

 Only elements that are 1 (or possibly 2 if they are larger elements) electrons(s) away from having complete valence shells are likely to become ions.

Let's introduce another rule of thumb that is a golden rule of chemistry.

 Opposite charges attract. Like charges repel.

The last rule of thumb is important because it explains another type of chemical bond. An **ionic bond** is the attraction between *two ions* of *opposite charges*. For example, Na^+ is attracted to Cl^- because they have opposite charges. The ionic bond holds sodium and chloride in close proximity to each other and creates NaCl, which is also known as table salt.

Unlike covalent bonds, there is no fancy *line model* that represents an ionic bond. Instead, we just write the element names next to each other, as demonstrated with NaCl.[2] Ionic bonds are not limited to two atoms; charged *molecules* can also form ionic bonds as long as the molecules have opposite charges. Ionic bonds between molecules will be important in future chapters.

> **Problem 7**
>
> Circle the three compounds that are held together by ionic bonds.
>
> a) NaCl   b) CaCl$_2$ c) NaK   d) H$_2$O   e) KCl

Alright! Let's revisit covalent bonds. Remember when I said that a covalent bond is like two siblings (atoms) sharing a shiny new toy?

Figure 5. Two seven-year-old children sharing their exciting new chemistry book equally (right), while a seven-year-old child able is able to take more ownership when sharing with a two-year-old (left).

I have two boys, a 7-year-old and a 2-year-old. A "shared" toy in my house means that the 7-year-old possesses it until he gets bored and puts it down. Only then can the 2-year-old sneak in and play with the toy (**Figure 5**). Likewise, when two atoms of different electronegativity share a pair of electrons, one of those atoms hogs the electrons more than 50% of the time. This is called a **polar covalent bond**. The word *polar* means that the electrons are polarized to one side in the covalent bond. The alternative is a **nonpolar covalent bond**, where the pair of electrons are shared equally between two atoms.

Are three little letters going through your head now - WTF? Let's simplify. Each covalent bond can be further categorized as either a *polar* or *nonpolar* covalent bond. Your job will be to determine whether a

---

[2] This notation can be confusing because it requires the reader know when they are looking at an ionic bond (NaCl) versus the chemical formula for something held together by a covalent bond (such as O$_2$ or H$_2$O). The charges can be written over the respective atoms or molecules such as Na$^+$Cl$^-$ to help the reader differentiate, but don't count on it.

covalent bond is polar or nonpolar. We will only concern ourselves with the four elements we see most commonly involved in covalent bonds – carbon, hydrogen, oxygen and nitrogen.

We need some more rules of thumb to help us determine when a covalent bond will be polar. Who has got two thumbs and wants to learn more chemistry?

Oxygen and nitrogen are "big sibling" elements.
Carbon and hydrogen are "little sibling" elements.

When the little sibling elements ask for their fair share of the electron orbits, think of the "big sibling" element saying "**NO!**" That will help you remember that **N**itrogen and **O**xygen (**N-O**) are the big siblings.

Here are the rules for determining if a covalent bond is polar or non-polar.

1. A covalent bond between a "big sibling" atom and a "little sibling" atom is **polar**.
2. A covalent bond between two "big siblings" OR between two "little sibling" atoms is **nonpolar**.

The following example demonstrates this. **Figure 6** shows methane with four covalent bonds. Each covalent bond is between a carbon atom and a hydrogen atom. Both carbon and hydrogen elements are *little siblings*. Therefore, the C-H covalent bonds are nonpolar because they are between two *little siblings* (see rule 2 above). Water has covalent bonds between an oxygen atom and a hydrogen atom. Oxygen is a *big sibling*; hydrogen is a *little sibling*. Therefore, a water molecule has two polar covalent bonds. Check your understanding of polar versus nonpolar covalent bonds with Problem 8.

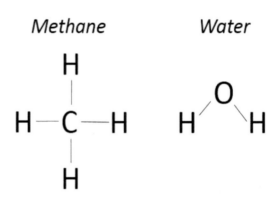

Figure 6. The covalent bonds in methane are nonpolar. Water has two polar covalent bonds.

Problem 8

Circle all polar covalent bonds in this molecule. Put a box around the nonpolar covalent bonds.

```
      H   H   O
      |   |   ||
   H—N—C—C
          |   \
          H    O—H
```

Finally, we explore the hydrogen bond – the most enigmatic bond of all! The name *"hydrogen bond"* looks deceptively simple. Yet the definition of a hydrogen bond is a paragraph filled with jargon. But here we will keep it simple. Our endgame is to predict when a hydrogen bond will form.

It all starts with a polar covalent bond. Remember that electrons are not shared equally in a polar covalent bond because the electrons spend most of their time around the "big sibling" atom (oxygen or nitrogen). This means that the "big sibling" atom adopts a partial negative charge. For a more detailed animation and explanation of this concept, place your smart phone or tablet over the following QR code.

In a water molecule, the two hydrogen atoms adopt a partially positive charge because the electrons are spending most of their time by the oxygen atom. Remember our golden rule of chemistry? Opposite charges attract! So, this partially positively charged hydrogen is attracted to other negative charges, but <u>not as strongly as two oppositely charged ions</u>.

If other polar molecules are nearby, something very interesting can result. The partially positive (δ^+) hydrogen becomes weakly attracted to a partially negative face of some other polar molecule. This weak interaction is a hydrogen bond (**Figure 7**). You can think of it as the electron-poor hydrogen atom essentially "window shopping" for another electron.

A hydrogen bond is longer and weaker than a covalent bond. A dashed or dotted line represents a hydrogen bond, and the use of a broken line is symbolic. After all, most covalent bonds are about 20 times stronger than hydrogen bonds, just like a solid line represents something sturdier than a dashed line.

In order for a hydrogen bond to exist, the following criteria must be met.

(1) You must have two molecules[3] each with a polar covalent bond.

(2) The "little sibling" for one polar covalent bond must be a hydrogen atom, which makes the hydrogen atom δ^+.

(3) The δ^+ hydrogen must be facing a δ^- big sibling (either an oxygen or nitrogen atom) on another molecule.

Figure 7 illustrates these three criteria are satisfied. We have two water molecules, each with polar covalent bonds. In all cases, the little sibling is hydrogen. The δ^+ hydrogen on the lower water molecule is facing a δ^- oxygen atom on another water molecule.

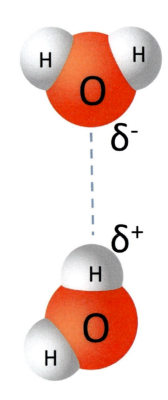

Figure 7. A hydrogen bond is represented as a dashed line between a partially positive hydrogen atom and a partially negative atom somewhere else.

---

[3] In cases where there is one very large molecule (such as protein), then a hydrogen bond is possible between separated regions of the large molecule that can wrap back around and interact.

Can a hydrogen bond exist between the two molecules shown in **Figure 8**? No! Apply the three criteria and you'll see that the first one is not satisfied. While we do have two molecules, the left molecule does NOT have a polar covalent bond! The molecule on the left has all nonpolar covalent bonds because we have all "little siblings" (carbon and hydrogen are both "little siblings"). Therefore, none of the hydrogens in the left molecule have a δ^+ charge.

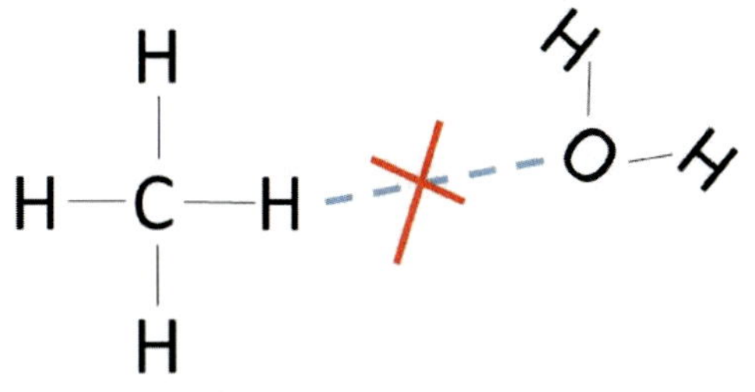

The best way to demonstrate your understanding is with another practice problem. Try out Problem 9, and make sure to check the answer key at the end of the chapter.

Figure 8. A hydrogen bond cannot exist between these two molecules because none of the hydrogens on methane (left molecule) have a partial positive charge.

Problem 9

Can you recognize a real hydrogen bond? Letters A through E mark the five chemical bonds to evaluate. Determine which of the lettered chemical bonds are true hydrogen bonds, and which ones are not. (*Hint:* Only two of the five bonds are hydrogen bonds).

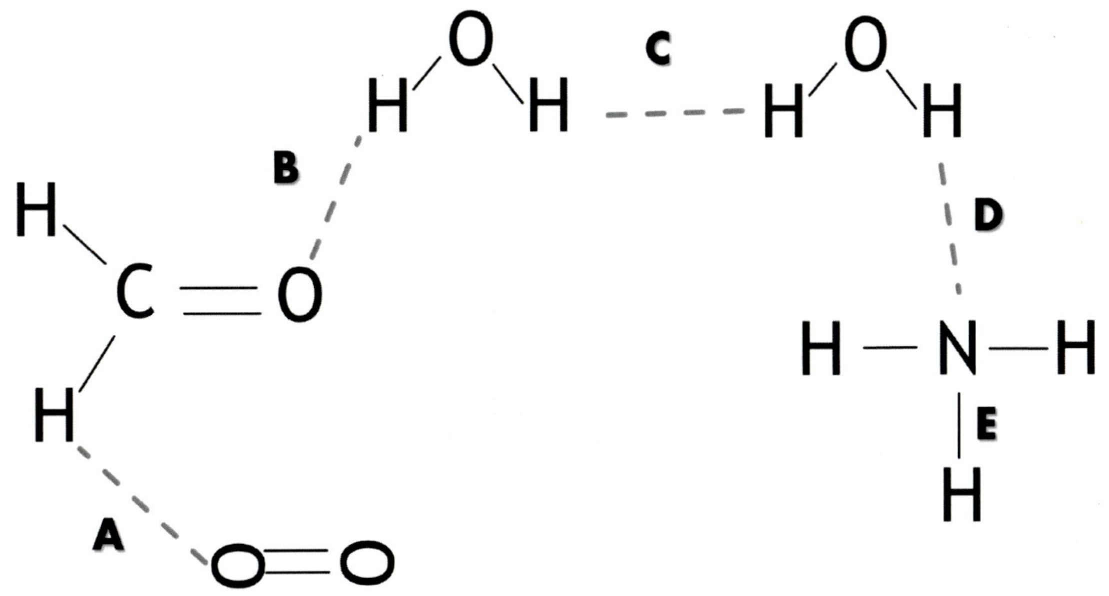

Congratulations! You made it to the end of the chapter. Before you go, try a few more problems just to be sure you've got a solid handle on the material. ☺

Problem 10

You are studying biology with a friend who draws out the following molecule. You notice that the molecule looks weird, so you take a closer look. Aha! Some atoms don't have enough covalent bonds, and some have too many. Identify all areas of the molecule that are incorrect.

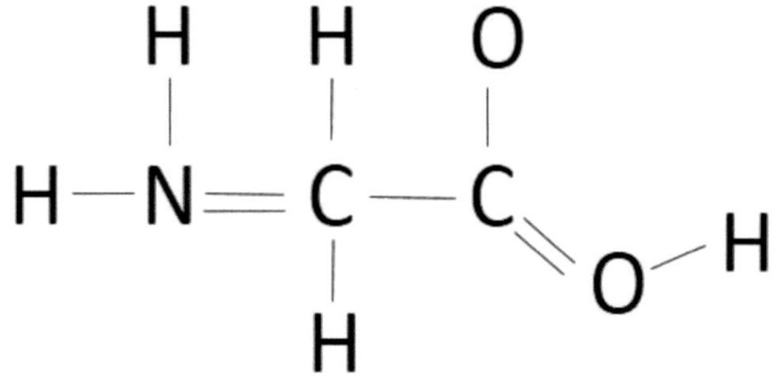

$$\begin{array}{c} H \quad H \quad O \\ | \quad | \quad | \\ H-N=C-C \\ | \quad \diagdown \\ H \quad O-H \end{array}$$

Problem 11

A. Circle all *polar* covalent bonds in each molecule.
B. Rank the molecules from most *nonpolar* to most *polar*, with #1 being the most nonpolar and #4 being the most polar.

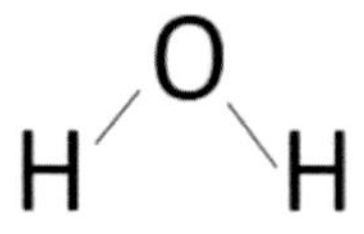

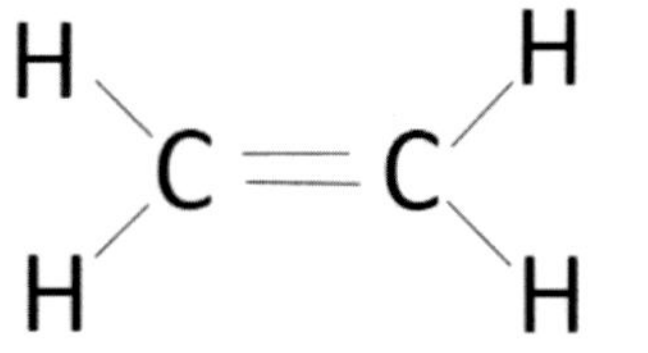

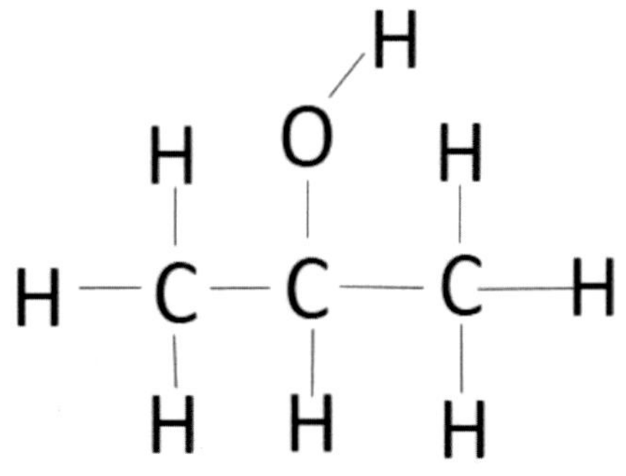

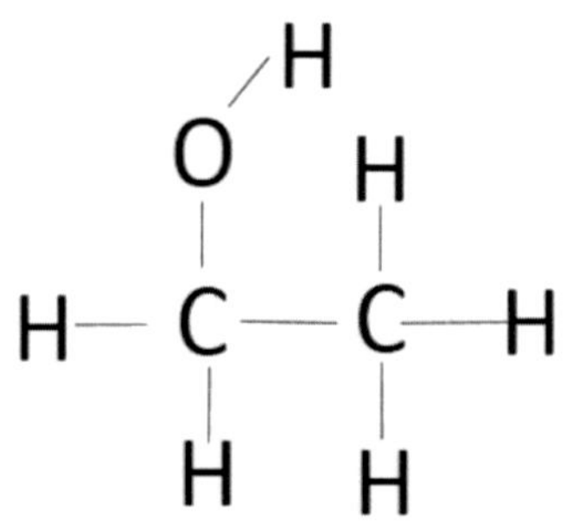

Molecule 1: H–O–H

Molecule 2: H₂C=CH₂ (ethene)

Molecule 3: H–C(H)(H)–C(OH)(H)–C(H)(H)–H (with H on the O)

Molecule 4: H–C(O-H)(H)–C(H)(H)–H

Chapter 2: Answer key

1. NH_3 is on the left; CH_4 is on the right. The blue electrons are from hydrogen; the grey electrons are from the central atom (nitrogen on the left or carbon on the right).

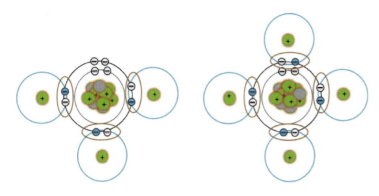

2. 2 covalent bonds; a total of 4 shared electrons. The following QR code links to a video with a more detailed explanation.

3. An oxygen atom forms 2 covalent bonds. A nitrogen atom forms 3 covalent bonds.

4.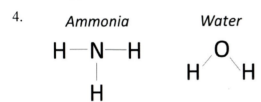

5. C

6. Calcium has 20 protons and therefore starts with 20 electrons. Calcium's ideal electron state would be 18 electrons because the third shell would be complete. Losing two electrons allows calcium to achieve a complete valence shell with 18 electrons. In this state, calcium still has 20 protons but only 18 electrons. This equates to a +2 charge (2 more protons than electrons).

7. A, B, E are all held together by ionic bonds. C is not possible as Na^+ and K^+ are repelled (like charges repel). D is water and is held together by covalent bonds, not ionic bonds.

8. All polar covalent bonds are circled. Nonpolar covalent bonds are boxed.

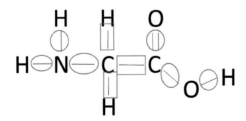

9. Answers B and D are the only true hydrogen bonds. Use your smart phone or tablet to access the following QR code for a more detailed explanation.

10. The red **X** represents an extra covalent bond that can't be there. A red line represents an additional covalent bond that is needed. If you are confused, use your smartphone or tablet to access the QR code for a more defined explanation of how to solve this problem.

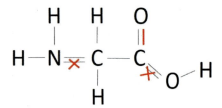

11. The molecules are ranked from *least polar* (in other words, *most nonpolar*) to the *most polar*. This is done by looking at % of covalent bonds that are polar. Ethene (#1) has no polar covalent bonds. Eighteen percent of isopropanol's bonds are polar, 25% of ethanol's bonds are polar, and 100% of water's bonds are polar. The fact that isopropanol is less polar than ethanol has implications as antimicrobial agents. Both ethanol and isoproponal are alcohols employed in hand sanitizers. Yet, one works at a lower concentration than the other because of slight differences in polarity. You may learn about these alcohols in microbiology.

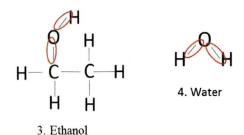

Chapter 3: Solutions

Back to the base-ics.

A good **solvent** is a solution that can dissolve many substances. From basic kitchen experience, you know that water is a very good solvent. Water dissolves many **solutes** (substances dissolved in a solvent), including sugar and salt, but it does not dissolve everything. Knowing what kinds of molecules are readily dissolved in water is helpful to understanding many human physiology and microbiology topics.

Water's capacity as a solvent starts with the fact that a water molecule is small but quite polar. Remember that opposite charges attract. Therefore, the partially positive hydrogen atom of a water molecule will be attracted to anything with a negative charge (partial *or* truly ionic). The oxygen atom in a water molecule carries a partial negative charge, and therefore is attracted to any positive charges on other ions or molecules. These attractions allows water molecules to surround charged ions or polar molecule and dissolve them. Place your smartphone or tablet over the following QR code to watch a video demonstrating how water can dissolve sodium chloride (NaCl salt).

Any substance that has an affinity for water is said to be **hydrophilic**. Not all substances are hydrophilic, though. For instance, oil molecules are not. Oils are molecules made up almost entirely of carbon and hydrogen atoms. Remember that carbon and hydrogen are both "little sibling" elements. A C-H covalent bond is nonpolar, which means no partial charges manifest. Consequently, we know that oil molecules are nonpolar. Moreover, experiences in a kitchen demonstrate that oil does not dissolve in water. In fact, oil and water repel each other. A **hydrophobic** substance is repelled by water. I bet you can come up with several hydrophobic cooking items beyond oils. Anything fatty also tends to be hydrophobic – peanut butter, grease, lard, ect. What do all of these hydrophobic substances have in common? They are almost entirely composed of carbon and hydrogen atoms (all "little siblings"), and therefore nonpolar. This rationale brings us to our next rules of thumb!

 Ions and polar molecules are usually hydro<u>philic</u>.

Nonpolar molecules are usually hydro<u>phobic</u>.

Although these rules of thumb give us a great starting point to predict how well something will dissolve in water, it is important to recognize that **the degree** to which a substance *"likes water"* is graded by charge. The more charged a substance is, the better it dissolves in water. The more nonpolar the molecule, the less it will dissolve in water.

Cooking dinner notwithstanding, why is this important again? **How well something dissolves in water correlates to how well it dissolves in your blood or in the cytosol (fluid) of your cell.** And how well solutes dissolve in your blood or cytosol is definitely important.

As much as I want to bombard you with physiological applications to this theme, I will limit our discussion to one fascinating example – the difference in solubility between O_2 and CO_2.

Figure 1 shows the line model for both O_2 and CO_2. Atmospheric oxygen is a completely nonpolar molecule. CO_2 has two polar covalent bonds <u>but is often considered to be a nonpolar molecule.</u> This seems like an oxymoron until you consider the symmetry of a CO_2 molecule. Place your smartphone or tablet over the on the left QR image to learn why CO_2 falls somewhere between the *polar* and *nonpolar* classifications. Then try Problem 1.

Figure 1. Carbon dioxide is slightly more polar than O_2.

Problem 1

A cup of water is exposed to 1000 gas molecules each of CO_2 and O_2. After 1 minute, there will be *more* _____ molecules dissolved in the water.

    a) *CO_2 than O_2*   b) *O_2 than CO_2*

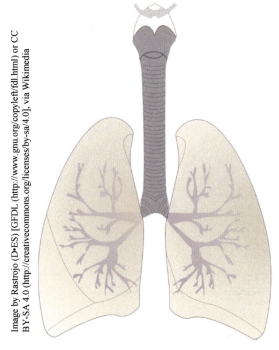

Figure 2. The lungs of our respiratory system are designed to bring in fresh O_2 from the environment and expel CO_2 from our bodies.

Our respiratory system (**Figure 2**) exchanges O_2 and CO_2 gasses between the air and blood. Yet, these two gasses are not equally soluble in water, and therefore not equally soluble in plasma, the fluid component of blood. **Remember - the more polar or charged you are, the better you dissolve.** At 37°C, a cup of water can hold approximately twenty times more CO_2 molecules than O_2 molecules simply because CO_2 is slightly polar and O_2 is not. *Why does this affect us?*

When we inhale, we bring millions of O_2 molecules into the deep recesses of our lungs and place them right next to our bloodstream. But O_2 is hardly soluble in the watery component of our blood (which is called *blood **plasma***) and so only a few of the O_2 molecules dissolve into our plasma. This is a real problem. As humans, we need much more O_2 than the tiny amount that dissolves into plasma. This very conundrum is why red blood cells are necessary for our survival. Red blood cells take up less than half of your blood volume, yet carry 98% of our oxygen due to their optimal design to capture and hold large quantities of O_2 (**Figure 3**). Put another way, without red blood cells, your blood would transport only about 5% of your O_2 requirement, leading to a quick death.

We can thank O₂'s hydrophobic nature and poor dissolution in water for that! In contrast, the majority of the CO₂ produced as waste **IS** transported in the plasma, in part because CO₂'s slightly polar nature means it dissolves better in aqueous fluids such as blood plasma.

Are your eyes glazing over? **Let's recap**.

- O_2 doesn't like water because O_2 is nonpolar and therefore hydrophobic.
- The watery component of blood, which is **plasma**, can't dissolve very much O_2.
- Red blood cells are the solution to the solution (that's a pun)! Red blood cells act as special taxicabs to capture O_2 from the lungs and hold it during transit in the blood.
- Most CO_2 travels in the plasma because CO_2 is slightly polar and therefore dissolves better in water.

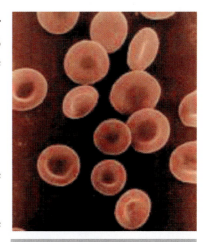

Figure 3. Red blood cells hold 97% of the O_2 in the blood. The liquid part of blood (plasma) holds almost no O_2 because O_2 dissolves poorly in water.

As you can see, how well a substance dissolves in water has profound implications on how living organisms are designed. Before we leave the topic of solubility, try Problem 2.

Problem 2

Rank the following chemicals from *least soluble* to *most soluble* in water.

NaCl

O=C=O

H-O-C-C-H (with H's on each carbon; ethanol)

H-O-C(=O)-O-H (carbonic acid)

O=O

29

 The rest of the chapter is dedicated to the pH scale, acids, bases, and buffers. **pH** stands for **p**ower of **H**ydrogen; it is a scale that reflects the concentration of H^+ ions in a solution. In a solution of pure water, the *concentration* of H^+ ions must **always equal** the *concentration* of OH^- (hydroxide) ions. Place your smart phone or tablet over this QR code for a quick lecture about why $[H^+] = [OH^-]$ in pure water.[1] This will be our starting point for discussing pH.

Think of the pH scale like a seesaw. The concentration of H^+ is at one end of the seesaw. At the other end of the seesaw is *a substance that removes* H^+, such as hydroxide ions (OH^-).[2] In a solution of water, the concentrations of H^+ and OH^-  are equal and the seesaw is perfectly level. Such a solution is *neutral* **(Figure 4)**. When the concentration of H^+ changes, the seesaw moves one way or another and the solution is no longer neutral. Any solution with a greater $[H^+]$ than pure water is an **acid**. A **basic** solution has a lower $[H^+]$ than water.

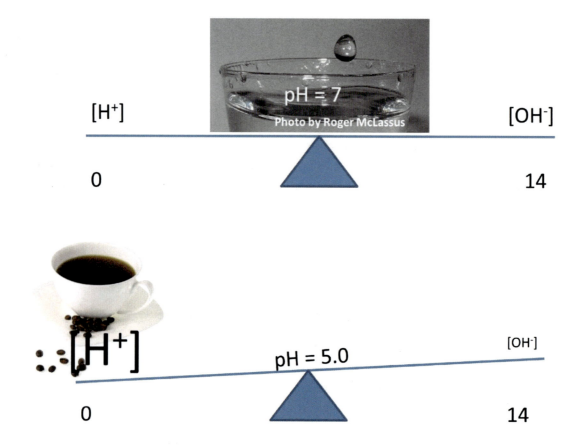

Figure 4. The seesaw analogy of pH. Underneath the seesaw is a numerical scale from 0 – 14. H^+ is on one end of the seesaw and a *substance removing* H^+ is on the other (OH^- is one of many substances that can remove H^+). The top picture shows pure water with a pH of 7. The lower picture is black coffee, which has about 100x more $[H^+]$ than water. Therefore, the seesaw dips to the left and the pH slides to a lower number.

---

[1] Square brackets represent concentration. So $[H^+]$ represents the concentration of H^+.
[2] Figure 4 shows $[OH^-]$ as the counterpart to H^+, but bases other than just OH^- that can remove an H^+ from solution and therefore raise the pH. For example, NH_3 is a base because it can suck up H^+ to form NH_4^+.

Because H⁺ concentrations are numerically messy, we've adopted the pH scale (ranging from 0 – 14) as a more user-friendly representation of "the **p**ower of **H**ydrogen (ion)". Unfortunately, the calculation of pH is not so user-friendly (unless, of course you just LOVE molarity and logarithms). In order to see the forest for the trees, we're going to bypass this messy math and explore the pH scale using the seesaw analogy as our starting point.

The pH scale ranges from 0 – 14[3]. The middle point in that range is 7 and therefore 7 represents a *neutral solution* such as pure water. An acid is any number lower than 7 (0 – 6.9) and a base is any number higher than 7 (7.1 – 14). The farther the pH is away from seven, the stronger the acid or base. That is, a solution with a pH of 2 is a <u>much</u> stronger acid than a solution with a pH of 6. How much stronger? Time for a *rule of thumb*.

 A pH change of 1 unit = a 10-fold change in H⁺ concentration.

This important rule can be clarified through some examples. Black coffee, with a pH of 5.0, has 100-times greater [H⁺] as water. How did we figure this out? On the pH scale, coffee (pH = 5) is 2 numbers away from water (pH = 7). Applying our rule of thumb, for each pH change of 1, there is a 10x difference in [H⁺]. With a pH change of 2, this is 10 x 10 = 100 fold difference in pH. And because a pH of 5 represents an acid, we know that this reflects a 100-fold *increase* in [H⁺].

Let's try another example. If we have a cup of seawater with a pH of 8, how does the H⁺ concentration compare to a cup of black coffee (pH of 5). The difference is 8 – 5 = 3. So seawater is 3 numbers away from coffee. This means 10 x 10 x 10 = 1000. But does seawater have 1000x *more* or 1000x *less* H⁺ than coffee? It would be less, because seawater pH is basic. Thus, the [H⁺] of seawater is <u>1000x less</u> than the [H⁺] of coffee.

Problem 3

Consider the pH value of the following solutions to answer questions A through D.

| Blood – 7.4 | Breast Milk – 7.2 | Stomach acid – 2 |
| Urine – 6 | Bleach – 12.6 | Milk of magnesia – 10 |

A. Which solution is the strongest acid? _____

B. Which solution is the strongest base? _____

C. Blood has _____ [H⁺] than breast milk.

    a) greater   b) lower

D. The concentration of H⁺ in milk of magnesia is _____ compared to stomach acid.

---

[3] The range of 0 – 14 is not arbitrary but rather a consequence of the way the pH scale is calculated using the logarithmic function. That explanation is beyond the scope of this tutorial.

We know that an acid is any substance that will increase the number of H⁺ ions in a solution, but just <u>what</u> are those substances? Acids are molecules willing to let an H⁺ go free in an aqueous solution. Hydrochloric acid is a great example of such a molecule.

Hydrochloric acid (HCl) starts with a chemical bond between chlorine and hydrogen.[4] In water, that bond is dissociated, and H⁺ and Cl⁻ are separated (**Figure 5**). In this way, adding HCl to an aqueous solution will increase the concentration of H⁺. Hydrochloric acid is important in human physiology as the acid produced by your stomach lining during digestion. During times of digestion, HCl secretion can bring the stomach juice pH as low as 1!

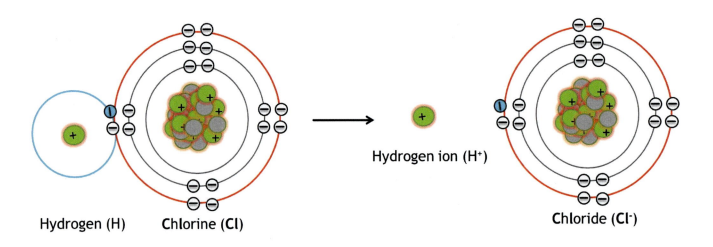

Figure 5. HCl readily dissociates into H⁺ and Cl⁻ in an aqueous solution, making hydrochloric acid a strong acid.

Many acids that you encounter in your study of biology are organic acids. Organic acids have one or more carbon atom(s) in their structure. Acetic acid (**Figure 6**) and amino acids are examples. What makes these molecules acidic is the **carboxyl group**, which is boxed in red in **Figure 6**. For reasons beyond the scope of this chapter, the carboxyl group's chemical arrangement allows it to easily release an H⁺ where the oxygen atom keeps the negative charge. Consequently, chemical structures with a – COOH group tend to be acidic.

On a side note......

Very strong acids and bases are highly corrosive. When stomach acid (which is a strong acid) is regurgitated up into the esophagus (the tube that connects your throat to the stomach) during indigestion or vomiting, it corrodes the tissue and causes a burning sensation. Because the esophagus is right behind the heart, this burning sensation is referred to as "heartburn" even though it is your esophagus, not your heart, that is being damaged!

---

[4] Figure 5 depicts the chemical bond holding hydrogen to chlorine as a covalent bond. In reality, the chemical bond holding the hydrogen and chlorine together is somewhere *in between* an ionic bond and a very, very polar covalent bond. Chlorine's electronegativity (pulling of electrons) is stronger than either oxygen or nitrogen.

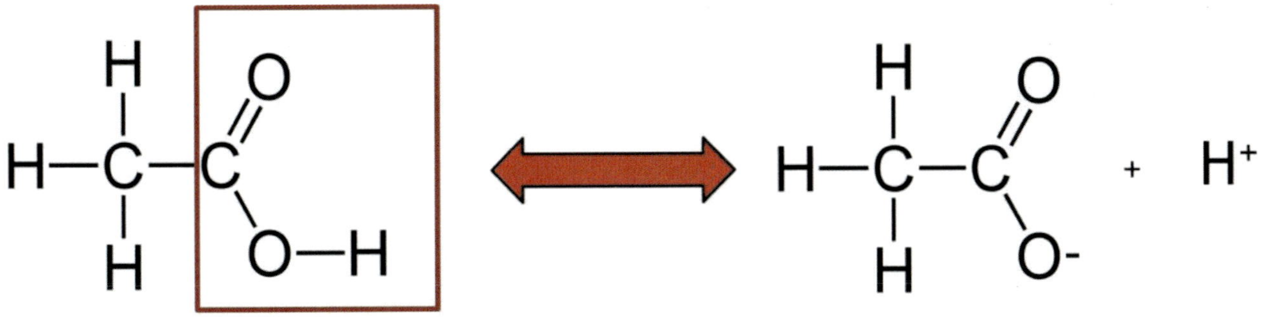

Acetic acid   Acetate

Figure 6. The chemical structure of the carboxyl group allows for an easy detachment of H^+ in an aqueous solution as shown with acetic acid. Once the acid releases the H^+ to the solution, the name of the molecule changes. The term 'acid' is removed and the other word incorporates the suffix "-ate". For example, acetic acid becomes acet**ate**.

Problem 4

Identify all organic acids from the structures below. Remember that an organic acid will have a carboxyl group.

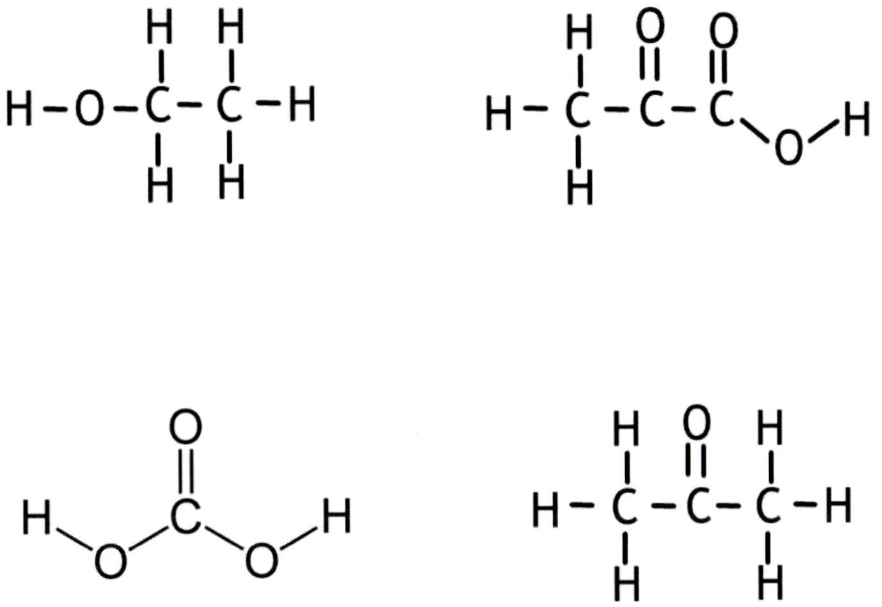

Let's have a similar discussion about bases. What substances are likely to act as a base? Recall that a base is any substance that *removes H^+ from a solution.* Let's start with sodium hydroxide. This very strong base serves as a classic example of how a substance can *remove H^+* from a solution.

Sodium hydroxide (NaOH) immediately dissociates into Na$^+$ and OH$^-$ in water. The hydroxide ion (OH$^-$) then quickly finds any free H$^+$ and forms water. Hence, the addition of NaOH to a solution effectively *removes H$^+$* from the solution as shown in **Figure 7**.

$$NaOH + H^+ \longrightarrow Na^+ + \underbrace{OH^- + H^+}_{H_2O}$$

Figure 7. Sodium hydroxide is a powerful base because it immediately dissociates into Na$^+$ and OH$^-$. The OH$^-$ then scrounges up a free H$^+$ to form water. In this way, NaOH removes H$^+$ from the solution.

While any substance that releases an OH$^-$ will act as a base, some chemical structures <u>with a nitrogen atom</u> are also primed to remove an H$^+$ ion from a solution. These are nitrogen bases, and the simplest of these is ammonia, which is NH$_3$. Understanding how a nitrogen base removes a proton is a bit tricky, so place your smart phone or tablet over the following QR code to watch how ammonia acts as a base.

Organic molecules with nitrogen atoms *often* act as weak bases. For example, an amine group is – NH$_2$ attached to a carbon atom; this group can act as a base in living systems. Again, it is nitrogen's unbonded pair of electrons that ropes in the hydrogen ion (H$^+$) with a covalent bond, and thus removes H$^+$ from solution.

Problems 5 & 6 bring us back to an important class of molecules that we've seen before, and we'll certainly see again – the **amino acid**. Amino acids serve as a great example of a molecule that acts as <u>both an acid and a base</u>.

Problem 5

This is glycine, one of 20 different amino acids. To familiarize yourself with the structure, please re-draw it in the white space.

Problem 6

A. Using the knowledge gained during the last two pages, (circle) the acidic part and put a [box] around the basic part of this molecule.

B. Draw the chemical structure of this molecule in an aqueous solution. Remember, the acidic part will give up an H^+ and the basic part will remove an H^+.

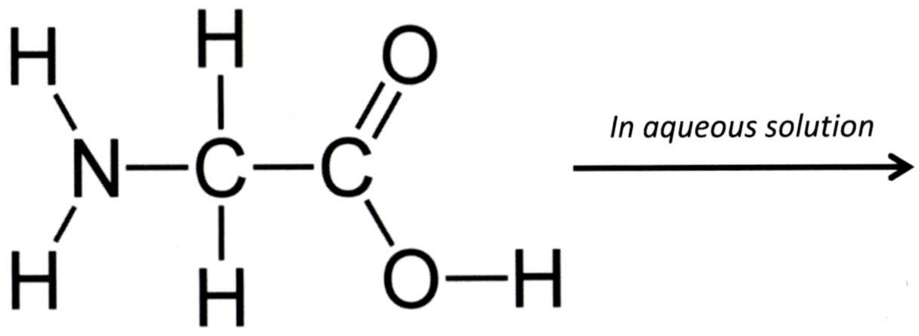

$$H_2N-CH_2-COOH \xrightarrow{\text{In aqueous solution}}$$

Check your answer for **Problem 6** before moving on to ensure a mastery of these concepts.

What happens when acids and bases are added together? The pH of the solution becomes more neutral as the H^+ (acid) and *substance removing* H^+ (base) are combined. Revisiting our heartburn example illustrates this nicely. Milk of magnesia is an over-the-counter solution that has several therapeutic uses. One application is to treat heartburn, which occurs when the corrosive stomach acid (HCl) gurgles back up into your esophagus and burns the lining. Milk of magnesia contains magnesium hydroxide, which is a base. When consumed, milk of magnesia helps neutralize the acid bubbling into the esophagus, thus minimizing further damage. Look carefully at the pH measurements taken before and after milk of magnesia is added to an acidic solution (lime juice) in **Figure 8**.

On a side note......

Mixing an acid and base together releases heat as the reaction is **exothermic**. The pH meter in **Figure 8** shows both the pH (top reading) and the temperature (bottom reading). Note that the temperature of the solution increased about 3°C when the acid and base were combined. Because highly exothermic reactions can cause the solution boil and even explode, you should never just "dump" a large volume of strong acids and strong bases together when trying to neutralize solutions.

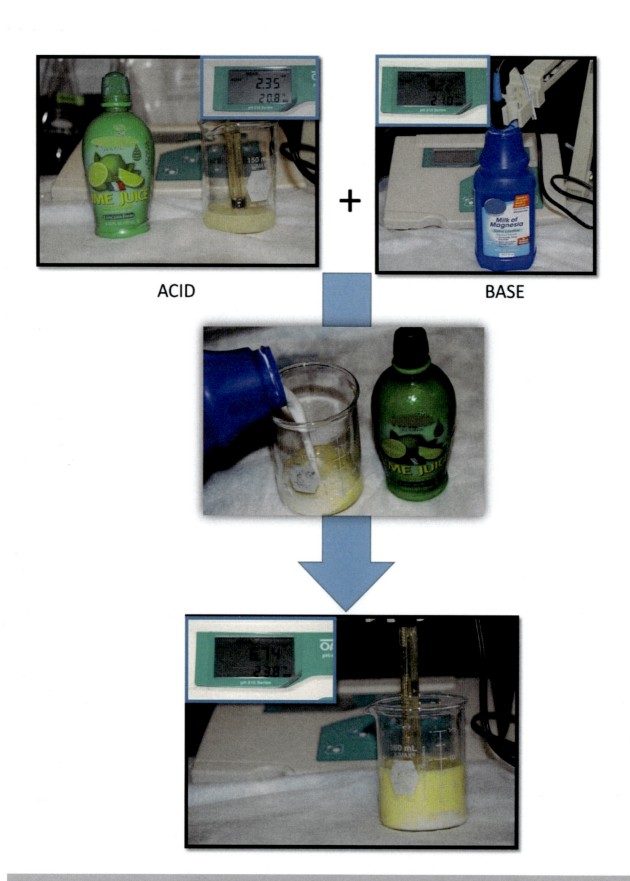

Figure 8. Milk of magnesia is basic with a pH of 9.2. Lime juice is acidic with a pH of 2.4. Combining the two brings the solution to a more neutral pH but also increases the temperature of the solution.

Problem 7

Hydrochloric acid (HCl) is produced in the stomach. Milk of magnesia contains the base magnesium hydroxide, which is Mg(OH)$_2$. In water, magnesium hydroxide dissociates into a single Mg^{2+} ion and two OH$^-$ ions. Write out the chemical reaction that occurs when magnesium hydroxide is added to stomach acid.

$$2HCl + Mg(OH)_2 \longrightarrow$$

Most living organisms can't tolerate huge changes in their pH. For example, the hardy *E. coli* bacteria survive best between a pH of 6 – 8 and can be killed in an acidic solution if the pH is below 4.0. Humans require a much more precise pH range. Our blood pH <u>must</u> stay between 7.35 – 7.45.

Because there is almost no margin of error for our blood pH, it's no surprise that our blood is full of **pH buffers** to help moderate the damaging effects of acids or bases entering into the blood plasma. **pH buffers** are exactly what the name implies – substances that "*buffer*" or dampen pH changes even when acids or bases are added to a solution.

On a side note......

Acidosis is when our blood pH drops below 7.35. A blood pH below 7.25 can be fatal. Acidosis is usually a symptom caused by a metabolic, respiratory or kidney abnormality.

When a person switches to a *no carb* diet, he or she risks developing **metabolic acidosis**. Metabolic acidosis can occur when our body's cells produce too many acids, and those acids get dumped into our blood faster than our kidneys can remove them. Unfortunately, breaking down large quantities of fats (which is what happens when a person deprives themselves of their usual energy source - carbohydrates) creates large quantities of an acidic waste product called "ketone bodies". These organic acids accumulate in the blood, which can lower the blood pH to dangerous levels.

pH buffers usually consist of one or more molecules that have the capacity to absorb an H$^+$ ion as the solution becomes more acidic, or donate an H$^+$ ion as the solution becomes more basic. That last sentence was a mouthful. Let's illustrate these features by looking at how bicarbonate works as a buffer. Bicarbonate is the major (though not only) pH buffer of our blood.

Place your smart phone or tablet over the following QR code to watch a demonstration of how bicarbonate helps buffer against pH changes when H$^+$ is added to or removed from a solution.

Problem 8

Draw out the chemical structure of bicarbonate *after* it reacts with NaOH, which is a strong base. (Revisit the previous tutorial on bicarbonate as needed).

Problem 9 *Challenge!*

Metabolic acidosis can cause the blood pH to drop to dangerous levels. Lactic acidosis is a *type* of metabolic acidosis, where excessive lactic acid builds up in the blood.

Assume you are a nurse caring for a patient with lactic acidosis whose blood pH has dropped to 7.25. Which might be the appropriate treatment(s) for this patient to bring the blood pH back up to 7.35? (Don't just guess or search the internet, try to reason out each answer).

a) a laxative to remove all of the lactic acid from the colon

b) Have the patient drink orange juice, which is acidic.

c) IV fluids containing bicarbonate

d) All of the above are appropriate treatments.

Chapter 3: Answer key

1. a) more CO_2 than O_2

2. From least soluble to most soluble with 1 = least soluble and 5 = most soluble.

   1. O=O

   2. O=C=O

   3. 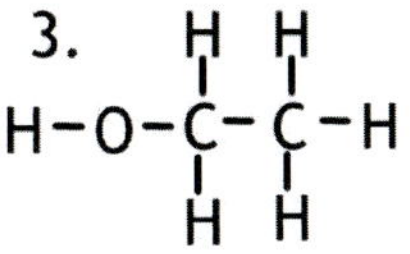 H-O-C(H)(H)-C(H)(H)-H

   4.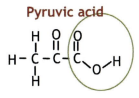

   5. NaCl

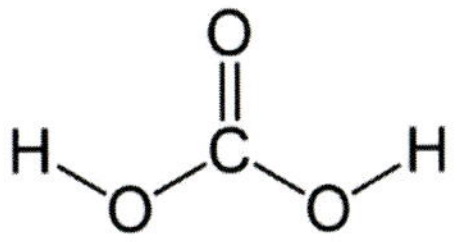

3. A. stomach acid   B. bleach   C. b) lower
   D. 10^8 or 100,000,000 times LESS

4. Pyruvic acid and carbonic acid are both acids (as their name describes). The component that makes each of these molecules acidic is the carboxyl group, which is circled in each molecule. Both ethanol and isopropanol lack the carboxyl group and therefore are not considered acids.

   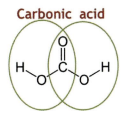

   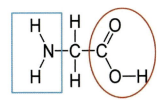

5.
6. A.

B. Place your smartphone or tablet over the following QR code to see how the amino acid looks in an aqueous solution.

7. $2H_2O + Mg^{2+}(Cl^-)_2$ If this was not your answer, please place your smartphone or tablet over the following QR code to listen to an explanation.

8. A strong base would *remove the H^+* from bicarbonate, leaving only carbonate behind. The chemical structure of carbonate is

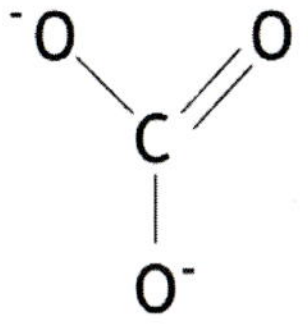

9. The correct answer is **C**. If this was not the answer that you got, please place your smartphone or tablet over the following QR code to listen to an explanation.

39

Chapter 4: Short-hand organic chemistry

Carbon is a promiscuous atom.

Organic chemistry is the study of molecules that have carbon atoms as their foundation. Organic molecules can be quite large, and therefore chemists have adopted a "stick figure" approach to sketch out organic molecules. This short-hand method is used instead of laboriously drawing out each atom.

For example, this molecule on the left is ethanol. We've seen ethanol in a previous chapter. The molecule on the right is also ethanol, according to chemistry short-hand.

If you're saying to yourself, *"Duh! Of course the molecule on the right is ethanol, I know how this works"*, then you may not need this chapter. However, if your thoughts are tending more towards the *"ummm.....what?"* line of thinking, then read on my friend. Learning this shorthand technique will not only help you immensely with future chapters, it's also kind of fun!

It all starts with something called a *carbon skeleton*. A carbon skeleton is a network of carbon atoms held together by covalent bonds, often in a chain or ring formation.

For example, if I ask you to draw a **chain** of four carbon atoms, I'm looking for something like this:

If I ask you to draw a **ring** of five carbon atoms, I'm looking for something like this:

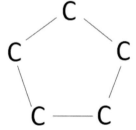

These chains or rings represent *carbon skeletons*, and they form the "backbone" (ha!) of the organic molecule.

Problem 1

On the left, draw a carbon skeleton with six carbon atoms in a *chain*. On the right, draw a carbon skeleton with six carbon atoms in a *ring*.

In Chapter 2, we learned that each carbon atom makes four covalent bonds. Therefore, the carbon atoms in the backbone or skeleton need additional atoms to partner up with! The easiest thing to do is "flesh out" (and the anatomy references/jokes just keep coming) our carbon skeleton with hydrogen atoms.

Place your smartphone or tablet over the following QR code to watch a tutorial on how to add the correct number of hydrogen atoms to a carbon skeleton. Then try Problem 2.

Problem 2

Finish the following carbon skeletons by adding the appropriate hydrogen atoms.

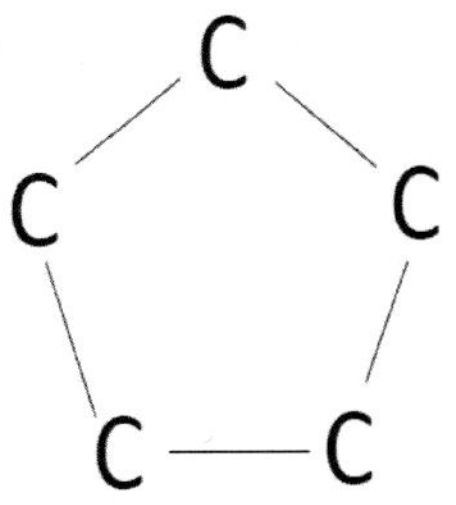

C—C—C—C

Problem 3

Are the molecules in Problem 2 *polar* or *nonpolar*?

Because each carbon can form up to four covalent bonds, carbon atoms can be versatile anchors for other atoms. Think of the carbon skeleton like a Mr. Potato Head; the hydrogen atoms are the standard face and outfit (eyes, nose, mouth, shoes and hat) that Mr. Potato Head comes with (**Figure 1**).

Just like you can remove and replace the facial parts and outfit pieces from a Mr. Potato Head, you can remove the hydrogen atoms from a carbon skeleton and replace them with other atomic arrangements, called **functional groups**. Where Mr. Potato Head parts plug into the potato body with a "pin", functional groups plug into a carbon skeleton with a covalent bond.

In biochemistry, we see the same atomic arrangements so often, we have names for them. The next few problems explore some common functional groups and their names.

Figure 1. Mr. Potato Head talking audio animatronic from a Disney park. Picture attributed to *Freddo* (Creative Common license by Wikimedia Commons).

Problem 4

Use an internet search engine or a biology textbook and draw out the chemical structures of each functional group. As an example, aldehyde has been done for you. (The red line for aldehyde is where that functional group attaches to a carbon atom).

Hydroxyl   Phosphate group   Ketone

Amino group   Aldehyde   Carboxyl group

$$\overset{\displaystyle O}{\underset{\displaystyle H}{\overset{\|}{-C}}}$$

42

Problem 5

The top molecule is a hydrocarbon composed of a six-carbon skeleton arranged in a chain. The bottom molecule shows that same molecule where some of the hydrogen atoms have been replaced by functional groups. In the bottom molecule, circle and label each of the following functional groups - the *hydroxyl group*, the *amino group*, the *ketone*, and the *carboxyl group*.

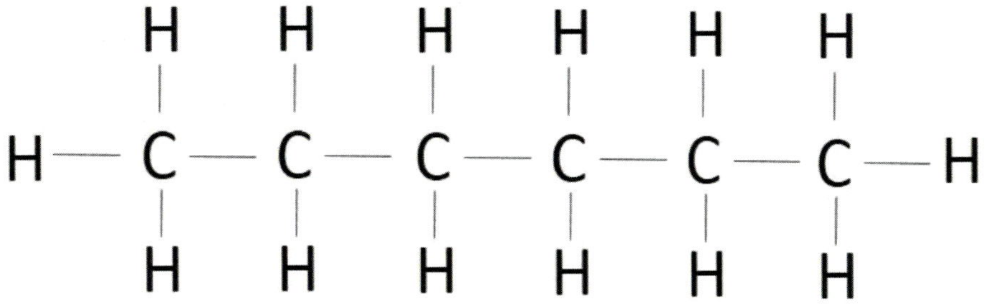

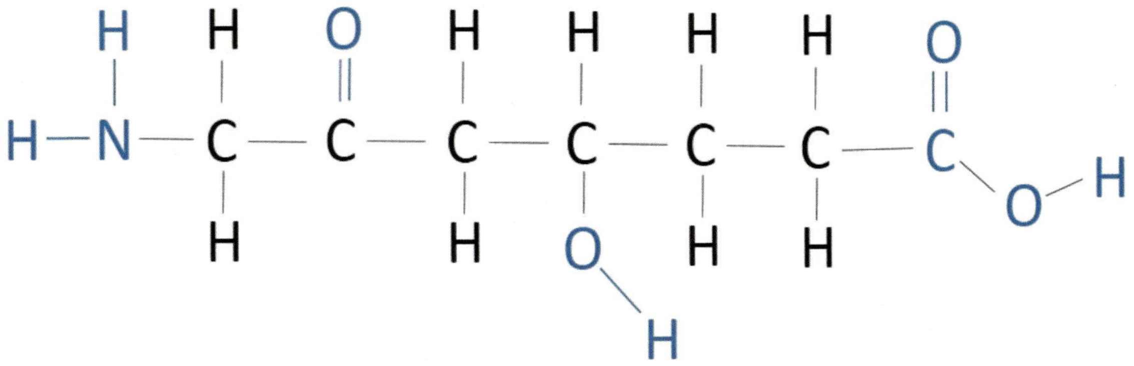

```
        H   H   H   H   H   H
        |   |   |   |   |   |
   H — C — C — C — C — C — C — H
        |   |   |   |   |   |
        H   H   H   H   H   H
```

```
       H   H   O   H   H   H   H   O
       |   |   ||  |   |   |   |   ||
  H — N — C — C — C — C — C — C — C
           |       |   |   |   |    \
           H       H   O   H   H     O — H
                       \
                        H
```

Problem 6

In Chapter 3, we learned that *ketone bodies* are molecules produced during fat metabolism. What type of functional group would you expect to be in ketone bodies? (Circle the correct structure).

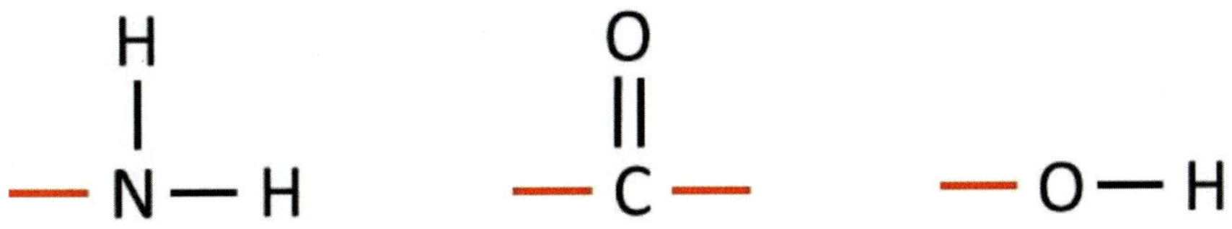

```
         H              O
         |              ||
    — N — H        — C —         — O — H
```

Chemists and biologists find it tedious to write out each atom for all but the smallest organic molecules. Therefore, chemists have come up with shorthand method of drawing organic structures. This "stick figure" approach saves much time in drawing organic molecules and most biologists get quite comfortable with the scheme. However, becoming comfortable requires a bit of time, patience and practice. The remainder of this chapter is learning how to alternate between long-hand and short-hand.

Figure 2 shows two ways to represent the organic molecule hexane. The left image shows all of the atoms, but drawing it out entirely is tedious. The right model is the 'stick figure', or shorthand, representation of hexane.

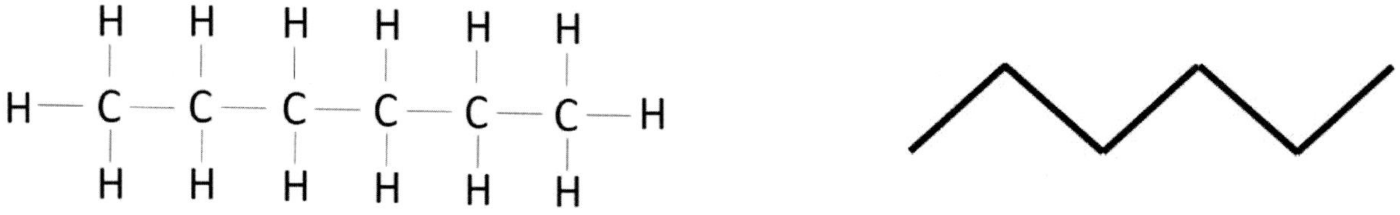

Figure 2. Both of these images represent the molecular structure of hexane, a hydrocarbon with six carbons in a chain.

Place your smartphone or tablet over the following QR code for a tutorial on why the Charlie Brown-esque zigzag pattern is an appropriate representation of hexane. Once you have reviewed the tutorial, try Problems 7 through 9.

Problem 7

Match each organic molecule in the first row to the correct shorthand molecule in the second row. Only three of the five shorthand molecules will match.

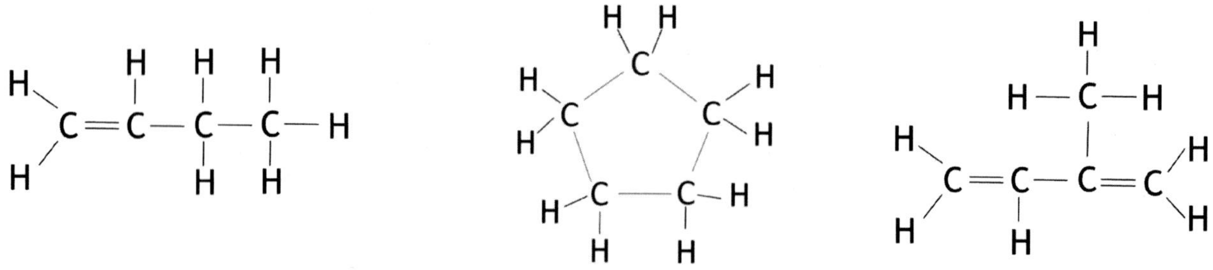

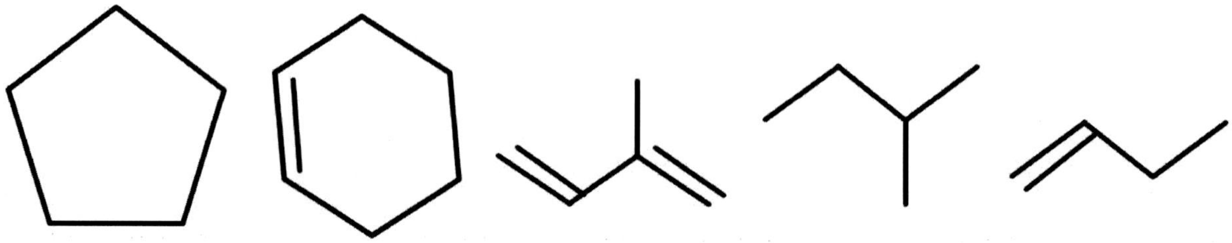

Check your answer for Problem 7 to ensure comprehension. Then try Problems 8 and 9.

Problem 8

Draw the shorthand structure for each organic molecule.

Problem 9

Draw the complete molecule for each shorthand structure listed below. Be sure to add in all appropriate hydrogen atoms.

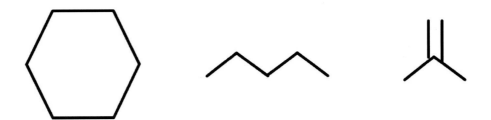

We're almost done! Now that you are comfortable with carbon skeletons and the shorthand method of depicting organic molecules, we can add in functional groups. Returning to our opening image, **Figure 3** shows how these are equivalent molecules. Notice that the functional group (hydroxyl) is still included in the shorthand structure.

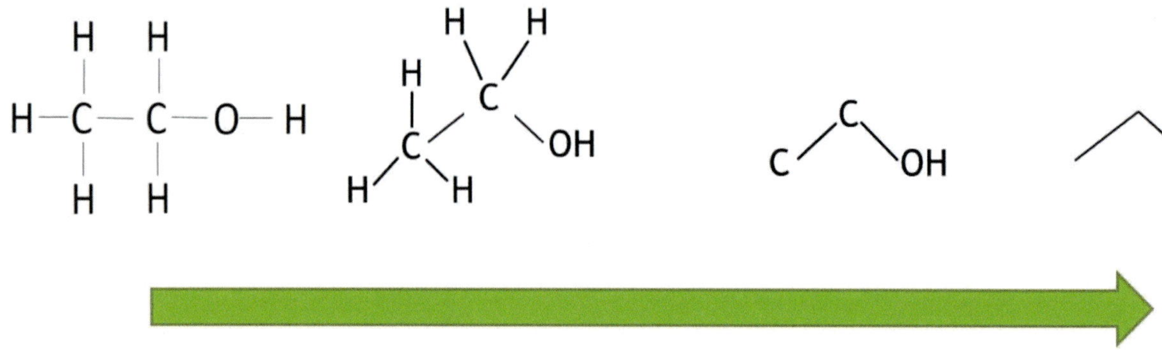

Figure 3. From left to right. The leftmost molecule shows all atoms of ethanol with a hydroxyl functional group. The next molecule also shows all atoms in ethanol, but the carbon skeleton has been bent in a manner consistent with the zigzag approach of the shorthand method. The third molecule shows ethanol without any of its hydrogen atoms attached to the carbon skeleton. (Because the hydroxyl functional group is not implied, it remains on the molecule.) The fourth molecule is the shorthand representation of ethanol. The rightmost line represents the covalent bond attaching the hydroxyl group to a carbon atom.

Problem 10 *Challenge!*

The top image shows both the shorthand (left) and longhand (right) image of an amino acid. The bottom image is a shorthand depiction of glucose (which is a sugar). Convert this shorthand drawing of glucose into the full molecule with all atoms shown.

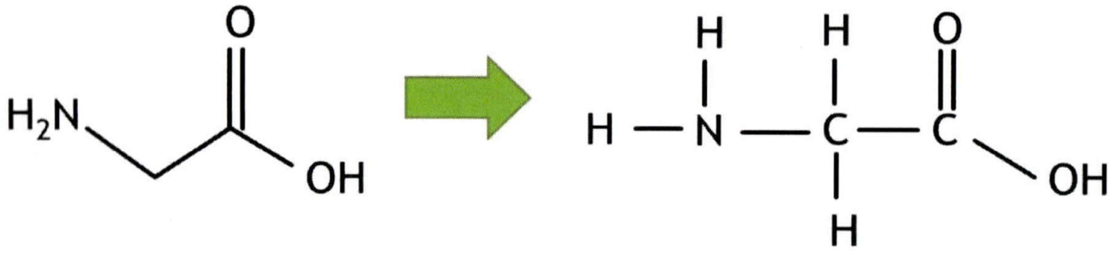

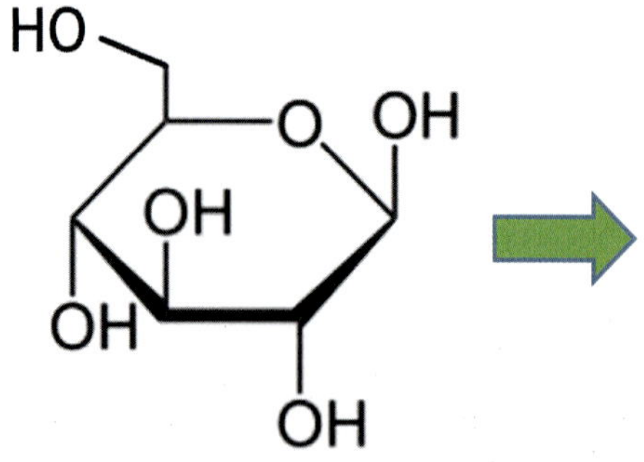

Chapter 4: Answer key

1.

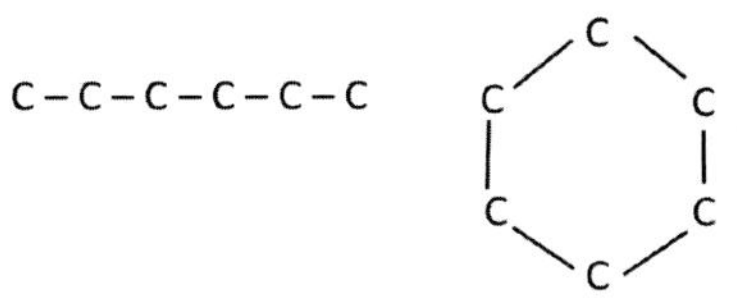

2. For an explanation on how to solve Problem 2, place your smart phone over the QR code below.

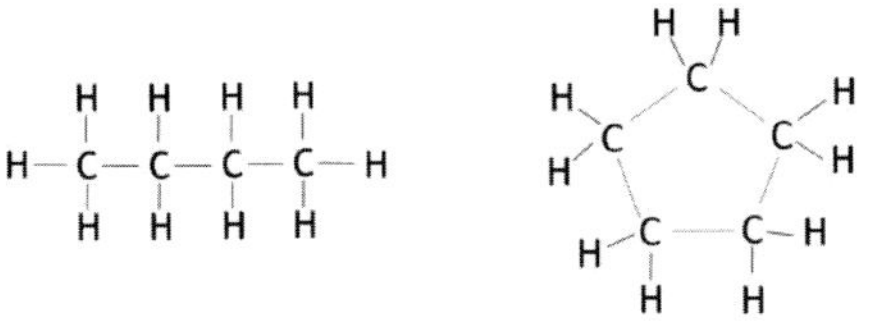

3. **Nonpolar** – the molecules have only non-polar covalent bonds (recall that both carbon and hydrogen are "little sibling" atoms).

4. Some functional groups are represented two ways. For example, hydroxyl groups are often written as (-OH) where the covalent bond between the oxygen and hydrogen atom is implied but not written.

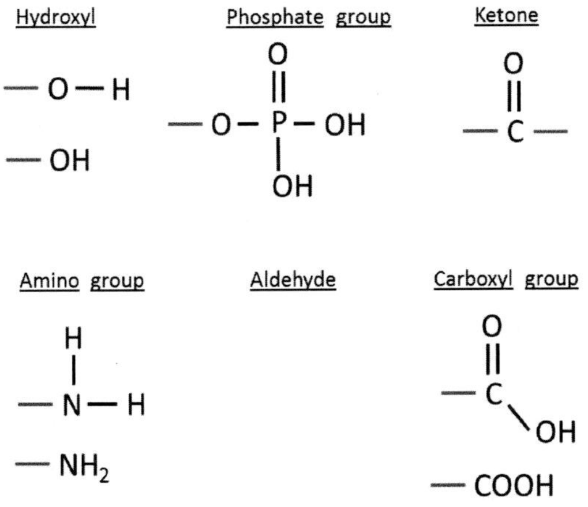

5.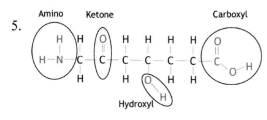

6.

7. The molecular structures on the left are matched to the shorthand structure on the right.

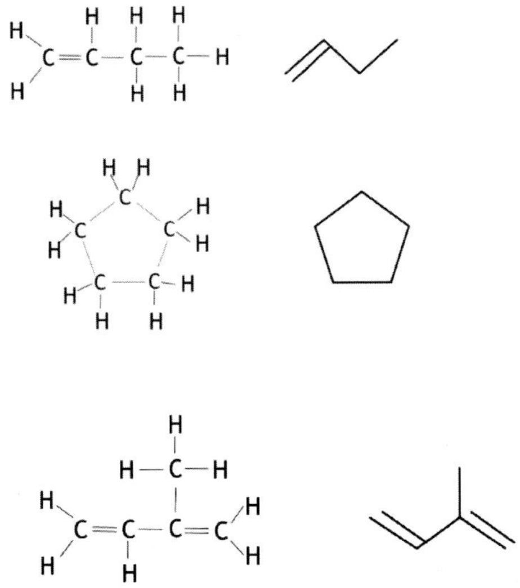

8. The following are the shorthand structures for the molecules shown in Problem 8.

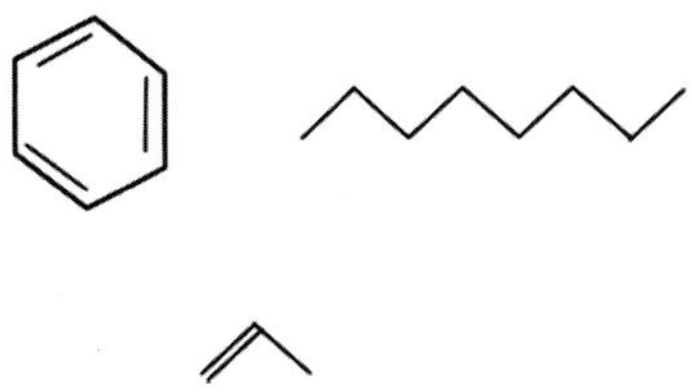

9. An explanation for Problem 9 can be found by placing your smartphone or tablet over this QR code.

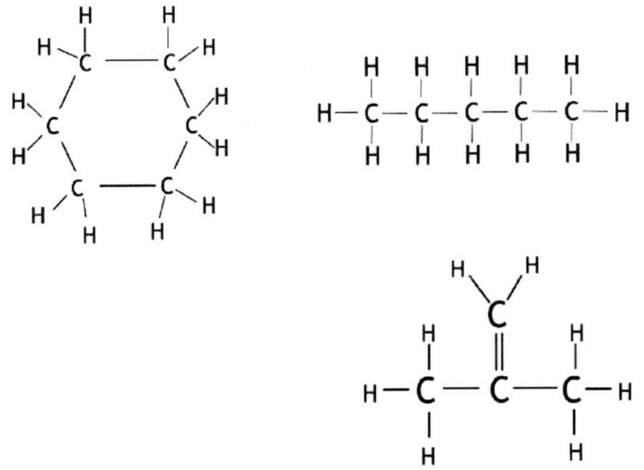

10. The answer and an explanation of the answer can be found accessing following QR code.

Chapter 5: Managing Human Anatomy

Work smarter, not harder.

The purpose of this chapter is to help you organize your studying approach to Human Anatomy. If Human Anatomy is your first course of the "big three" (i.e. Human Anatomy, Human Physiology and Microbiology), then this chapter is definitely for you! If you've already mastered Human Physiology and/or Microbiology, it's probably still worthwhile to peruse this chapter.

Reflect on all of the college courses you have taken thus far and organize them based on how much of the material was *conceptual* or *applied* versus *direct memorization*. For example, an art class or fitness class tends to be applied. You probably wouldn't make stacks and stacks of flashcards for such classes. Conversely, a class like Spanish I or American History requires lots of memorization; for such classes, flashcards might be your most effective study tool.

I've organized the scope of Human Anatomy, Human Physiology and Microbiology courses in a similar fashion. Students are often surprised to find that Human Anatomy and Human Physiology are on opposite ends of the spectrum, and this can explain why students sometimes ace one course and struggle in the other.

Human Anatomy is a very content-intensive course; you will be memorizing at least 2000 (!!!) new terms and structures. Taken individually, most terms and anatomical structures are straightforward, but memorizing several thousand of them in 16 weeks or less is very challenging. In contrast, Human Physiology explores the process of body functions and therefore is more conceptual. For example, learning how the nervous system conducts electrical information is conceptual and must be thoroughly understood, not just memorized. Consequently, any student who hopes to "flashcard" (i.e. memorize without necessarily understanding) their way through Human Physiology probably won't be earning a top grade. Microbiology is really a different type of class, and generally falls in middle on the spectrum. Instructor style can heavily skew your experience one direction or the other. For instance, I tend to favor process over memorization, so my Microbiology course is somewhere in the turquoise area. Some instructors teach Microbiology more in the yellow area. One challenge that is exemplified in Microbiology is that the topics covered are not visible to the naked eye. Getting comfortable with the *invisible* will be instrumental to your success in that class.

Since Human Anatomy has more memorization than any other class, and therefore the most information to manage, let's turn our focus to this class exclusively. Here is our first rule of thumb.

Managing large quantities of information in a short amount of time is the key to success in *Human Anatomy*.

Topics in Human Anatomy are not hard when approached individually – most 5th graders could learn what you are learning. Yet there are more failing grades in Human Anatomy than either Human Physiology or Microbiology. Chew on that for a moment. No one part of Human Anatomy is especially difficult, yet up to 50% of students fail or drop. What separates the successful students from the rest? Those who ace Human Anatomy successfully manage and process the huge *quantities* of information efficiently.

A major objective of this chapter is to discuss strategies to get the most learning done with the study time you have. Rather than *tell* you what will work, I will **show** you by conducting a personal experiment.

Assume that you are in a Human Anatomy laboratory and you are tasked with learning all of the structures of the ear. You may be provided with a list similar to the one below. You will be tested on this material in an upcoming laboratory practicum.

- Auricle
- External acoustic meatus
- Tympanic membrane
- Auditory ossicles (malleus, incus, stapes)
- Cochlea
- Eustachian tube
- Utricle
- Vestibule
- Saccule
- Semicircular canals and ducts (anterior, posterior, lateral)
- Tensor tympanii
- Stapedius muscle
- Round window
- Oval window
- Perilymph
- Endolymph
- Bony labyrinth
- Membranous labyrinth
- Vestibular duct
- Cochlear duct
- Tympanic duct
- Organ of Corti
- Cranial nerve VIII (Vestibulocochlear nerve)

What are you going to do with that list?

That's the million dollar question.

Before answering, digest this factoid. The average Human Anatomy course covers somewhere around 2500 structures in about 25 class sessions. That means that you will be learning an average of 100 structures PER class session. This ear anatomy list is only about 25 structures, or ¼ of what you will need to learn that day. Mathematically, this translates to 1 hour or less of class time to learn this list. Perhaps it is evident now why so many students fail.

Your goal or objective is to ace the exam. This means you need to know *how* you will be tested on this information. In the laboratory component of the course, exams are usually organized as a **practicum** (or **practical**). A typical practicum might be designed like this – the classroom has various human body parts spread out through the room as *stations*. For example, I usually have between 20 and 30 stations for a 100-question practicum. Associated with each station are a few questions about the model or specimen on display at that particular station (**Figure 1**).

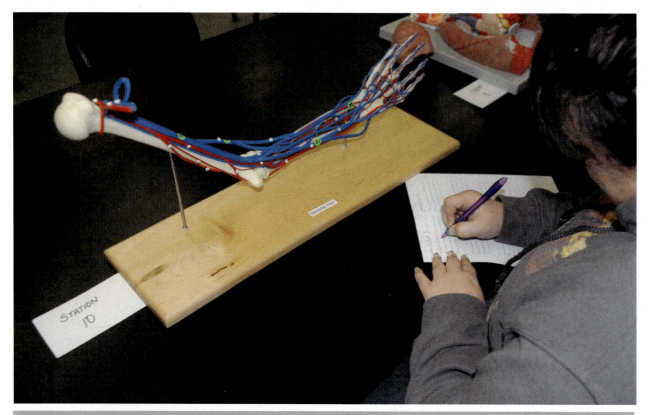

Figure 1. A student is asked to identify the blood vessels of the arm marked with green stickers by writing out their names on the exam sheet. This is a typical station in a laboratory practicum.

This part is important. Most of the practical questions require that you spell out the name of a marked structure on the model or specimen. **Being forced to write out the name of a structure in an exam is distinctly different than a multiple choice format;** in multiple choice, you simply need to *recognize* a term from a short list. And although a few instructors are lax on this, most instructors demand perfect spelling on their exams, and for good reason. After all, writing that a patient is suffering from a torn *labium*, when you mean torn *labrum*, on a patient's chart would cause inexcusable confusion and embarrassment.[1]

Thus, your endgame is to be able to achieve these three objectives.

(1) **Identify the specific anatomical structures on models and/or specimens.**

(2) **Recall the name of that structure.**

(3) **Spell the term correctly.**

---

[1] A torn labrum means the cartilage ring that reinforces the hip or shoulder socket is split. Torn labrums are common injuries. A torn labium basically means a damaged or injured vagina.

And for at least 2000 structures, if not more.

Take a second look at the ear anatomy list. It should be evident that you have a daunting task to complete in under an hour. Most students at this point would want to look at a physical model of an ear (preferably labeled), but there may only be a few ear models for the entire class. So, chances are you will have limited time with any physical model. Thus, it's no surprise that many students first find a 2D image to study, such as this one on the right (**Figure 2**). As you look at **Figure 2**, you should be able to cross a few structures of your ear list. Namely, the *saccule*, *vestibule*, *utricle*, *cochlea*, and *semicircular canals* (anterior, posterior and lateral). But unless you have previous knowledge on the inner ear, most of these terms and structures are probably unfamiliar. In fact, you may not even have a frame of reference for that image (*it came from the ear, you say? I thought it was a squid*). Coping with such unfamiliarity isn't the only battle. Perhaps you've noticed that exactly what the labels are referring to isn't always clear. For example, what's the difference between the *utricle*, the *vestibule*, and the *saccule*? That's not clear in **Figure 2**.

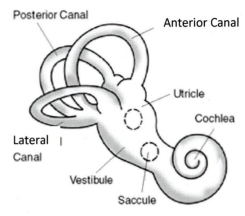

Figure 2. A diagram of the inner ear.

A close look at another perspective, such **Figure 3**, helps clarify it. The *utricle* and *saccule* appear to be distinctive sacs located within the *vestibule*. But it takes considerable research – thumbing through your text, using an internet search engine to uncover additional images, or asking your instructor – to figure that out. And all of that research takes time. For many students, most of the laboratory period is allocated trying to figure out just *what* is *what*.

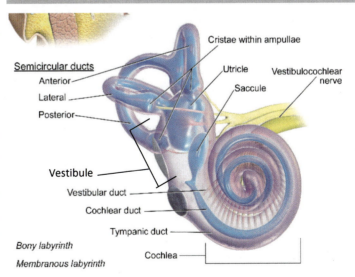

Figure 3. A more descriptive and detailed image of the inner ear.

Image by Blausen.com staff. "Blausen gallery 2014". Wikiversity Journal of Medicine. DOI:10.15347/wjm/2014.010. ISSN 20018762. (Own work) Creative Commons license via Wikimedia Commons.

So, why do so many students fail?

Because the scenario I just described only covers the first of the three objectives. And the fatal mistake so many students make is they continue to study in a manner that only masters that first objective. Such an approach might work for multiple choice-style exams, but you need to master all three objectives to ace a practicum.

Let's explore three different studying techniques through a personal experiment. The following pages will show you a detailed, labeled illustration of a particular human anatomical structure. **You will be directed to study the anatomy using a prescribed technique.** You will then be tested on how many structures you can recall. The goal is here is **metacognition** – to be cognitive (or aware) about how to best learn this material. To make this section work for you, follow the directions carefully and precisely.

TECHNIQUE 1 – *Reading*

Set a timer for five minutes and study the following image. Look at it carefully, study each word, but do not write down or say any word out loud. Just **read** the words and try to memorize them. After five minutes, turn the page.

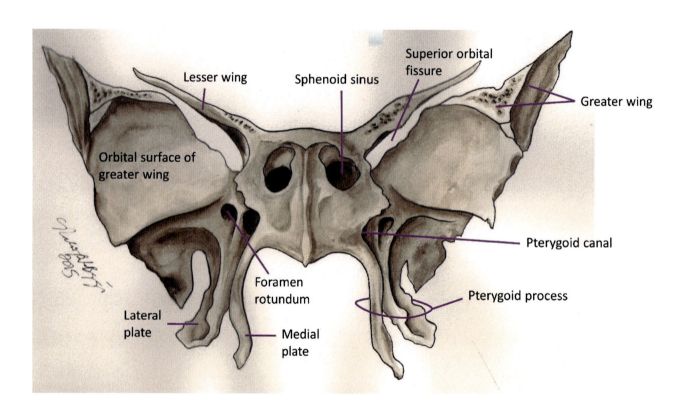

THIS PAGE IS INTENTIONALLY BLANK

TECHNIQUE 2 – *Writing*

Set a timer for five minutes and study the image on the left, which shows the underside of the brain. The yellow lines are nerves. There are 12 cranial nerves. Look at nerves and their names carefully and study each word.

At the end of five minutes, you will label all of the cranial nerves in the right-hand image, looking at the labeled image as needed to recall names and spelling. Because cranial nerves are paired, you can write down the names of each nerve twice (once on the left and once on the right). Take as much time you need.

When you finished labeling, turn the page.

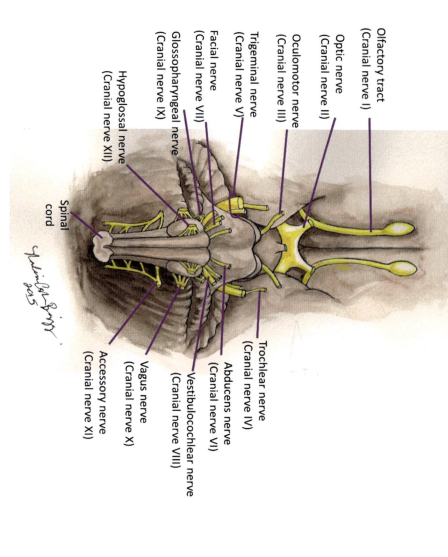

Olfactory tract (Cranial nerve I)
Optic nerve (Cranial nerve II)
Oculomotor nerve (Cranial nerve III)
Trigeminal nerve (Cranial nerve V)
Facial nerve (Cranial nerve VII)
Glossopharyngeal nerve (Cranial nerve IX)
Hypoglossal nerve (Cranial nerve XII)
Spinal cord
Accessory nerve (Cranial nerve XI)
Vagus nerve (Cranial nerve X)
Vestibulocochlear nerve (Cranial nerve VIII)
Abducens nerve (Cranial nerve VI)
Trochlear nerve (Cranial nerve IV)

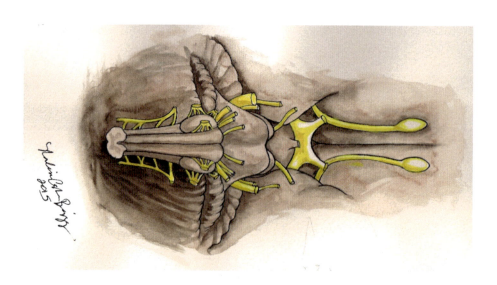

THIS PAGE IS INTENTIONALLY BLANK

TECHNIQUE 3 – *Drawing, labeling and pronouncing*

Examine the labeled image of the brain stem. Then draw (or copy) the brain stem in the box. If you are not a good artist, don't worry – use geometric shapes (circles, lines, ovals, ect.). The only requirement is that you take at least five minutes drawing the image. Once you have sketched the image, you may color or shade it if you wish. Then label each structure on your drawing. Once you have labeled it, pronounce each word to the best of your ability. Use your smartphone or tablet to access the QR code on the right to help you with the pronunciation. Take as much time as you need.

When you are finished drawing, labeling and pronouncing each word, read the directions on page 57.

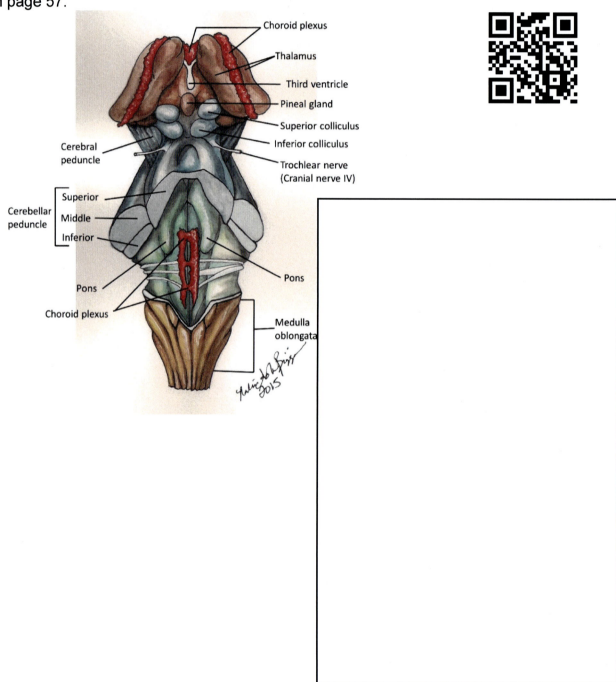

TAKE A BREAK FOR AT LEAST 20 MINUTES.

IT IS CRITICAL THAT YOU DO NOT STUDY DURING YOUR BREAK.

WHEN YOU RETURN AFTER YOUR 20 MINUTE BREAK, SET A **TIMER FOR 10 MINUTES** AND TURN THE PAGE.

Identify as much as you can from memory in ten minutes. Write down the name of each structure marked by a black or purple line. It is very important that you do **NOT** go back to look at any labeled images for review. *The goal is to test yourself as if you were in an Anatomy laboratory practicum and asked to identify all of these structures.* Try your best.

This is the _____ bone.

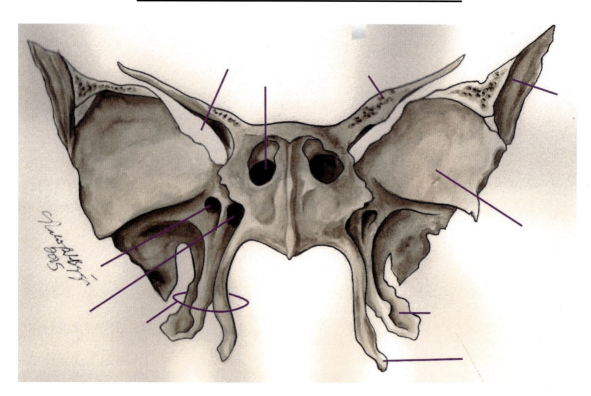

Once you have identified as much as you can remember, go to page 59. You may flip back and forth between these two pages until the 10 minute timer expires.

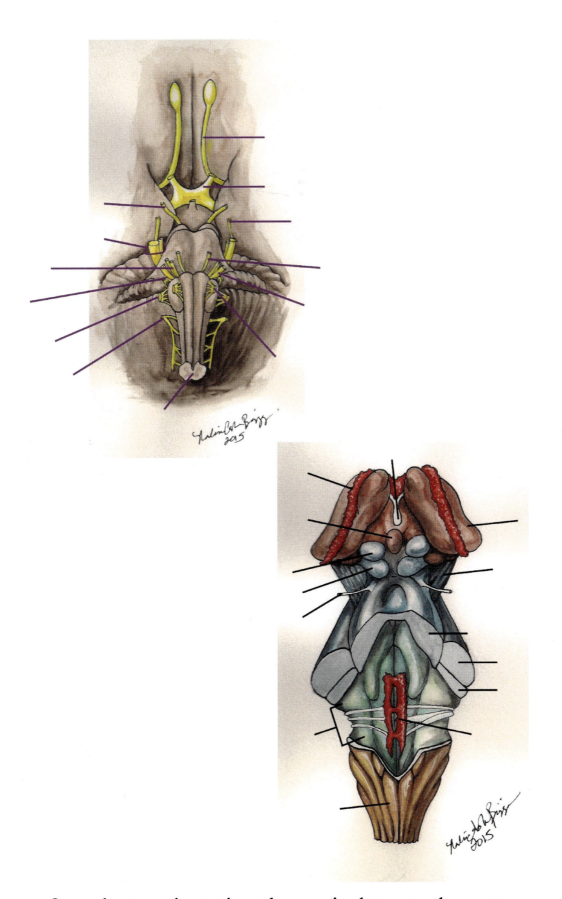

Once the ten minute timer has expired, turn to the next page.

At this point, you will score your answers from the previous page. Score your points honestly.[2]

Give yourself one full point for correct identification and perfect spelling. For answers that were *close* and had no more than two **letters** incorrect in your spelling error, award yourself 0.5 points. Any answer that has two or more letters incorrect (spelling errors) or incorrect identification earns zero points.

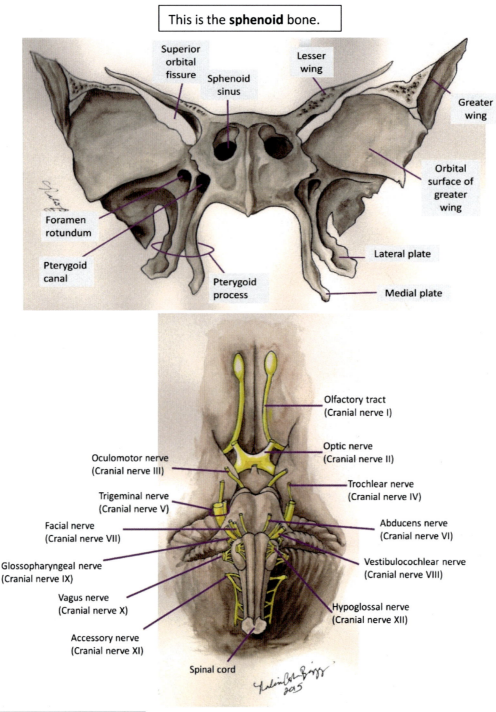

---

[2] Scoring honestly is in your best interest. If you score generously to make your score higher here, the only person you are cheating is yourself. Don't expect your instructor to give you the same "benefit of the doubt".

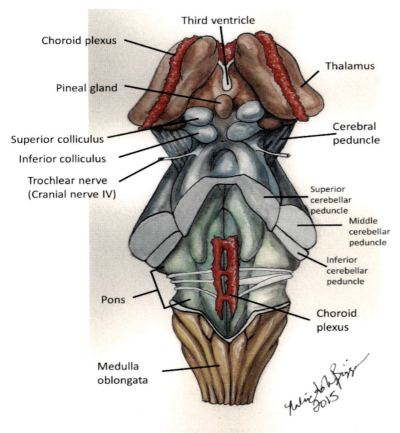

Hopefully this experiment has shown that not all study techniques have equal value. Many students use the first technique exclusively. And it is difficult to commit large volumes of information to long-term memory through this *reading-style* approach only. Indeed, one of the classical learning studies (the learning pyramid) found that only 10% of material is retained in long-term memory with Technique 1.

Technique 2 is an improvement. Spelling out each word twice by carefully labeling has the added benefit of forcing you to pay attention to the spelling. Technique 2 better prepares you to answer practicum questions, as your results hopefully demonstrated.

Technique 3 is the superior technique. Biology instructors have long recognized that drawing an object in detail induces you to study the object closely and analyze it carefully. You are much more likely to find patterns and relationships when you carefully draw an object. Moreover, orally pronouncing the word aloud is another method of forcing you to pay attention to the word and thus help commit it to long-term memory. In one of my Human Physiology courses, a clever group of students experimentally showed that ability to recall of a list of words increased when students were forced to say the words on the list aloud, as opposed to just reading the list.

It is probably not lost on you that Technique 3 is the most time-intensive. Time itself is an important factor in your success. As I will discuss shortly, putting in enough time is necessary to succeed in Human Anatomy.

But first, let's talk about how to implement Technique 3 in the most optimum and efficient manner. It's time for another rule of thumb.

 The most opportunity for learning occurs during *Human Anatomy* lecture and laboratory.

In other words, you need to use your time during class wisely and try to get the most learning done during your allotted class time. This means you should NEVER arrive to lecture and lab ice cold (i.e.

without any prior exposure to the material). Consider how you feel about a math lecture when you are behind on your homework. Because you have not mastered the previous material, that math lecture probably feels awkward and confusing. In fact, you may be totally lost. Now consider the anatomy lecture in a similar light. If you arrive to lecture and you have not even glanced at the material in advance, you'll spend most of your class time trying desperately to become "un"-lost. Consequently, you will have missed out on the most valuable learning opportunity for that material.

So what should you do? **Always start the class session with some working knowledge of the material.** You will absorb and learn much, much more during that class time. The day before the class session, allocate 45 minutes to previewing the material. This includes reading (skimming at least) the text for lecture material, and reviewing lab lists such as the one presented on the ear on page 49. Try Technique 3 in advance of lab – draw out an image from your text or other resource, label the structures, and orally pronounce the words. Try Technique 3 for each anatomical system to be covered the next day. You'll be amazed at how much more learning you will accomplish when you step into class. In lecture, your comprehension will skyrocket – you'll be nodding along eagerly while others are nodding off in confusion. In lab, the 3D models and real specimens will be illuminating, offering great opportunities for clarity and understanding by providing perspective that 2D pictures lack.

In summary, your capacity to understand and absorb the material in class, with all of its resources, will increase dramatically if you take the time to work with the material in advance.

Once you leave class, you should repeat the process outlined by Technique 3, drawing the same structure from different angles if possible. A great self-assessment test is to get an unlabeled image from the internet, print it out, and label it as best you can without any notes. You are well prepared if you can identify and label everything the instructor asks for, with 100% correct spelling, without any outside resources. If not, you need to keep drilling.

A final note. I emphasize that reading this chapter won't guarantee you an "A" in *Human Anatomy*. Even the most efficient students STILL require all of the allocated classroom time and about 15 quality hours outside of the classroom studying (not including food/bathroom breaks) per week. The 15 hours outside of the classroom per week assumes a full-semester 16 week course. It would take 30 hours/week in an 8 week summer course.

Did you process that last paragraph? If not, let me speak plainly. Acing *Human Anatomy* still requires a huge time commitment on your part. And there is no magic pill that can be substituted for that time. The techniques presented in this chapter can help you optimize that time, but are not a substitute for that time. And if you don't have the time this semester, your best option is to consider taking it another semester.

Chapter 6: Carbohydrates and nucleic acids

Organic chemistry at its finest.

Most organic molecules in living systems belong to one of these four categories – carbohydrates, proteins, nucleic acids or lipids. Carbohydrates, proteins, and nucleic acids start with a small molecule called a **monomer**. Monomers can exist individually, or they can be linked together by covalent bonds to form chains. These chains are called **polymers**. An analogy would be to think of a monomer like a bead (**Figure 1**). A polymer is the necklace of beads linked together.

As we review the four classes of biomolecules, the *monomer - polymer* theme should become more apparent. This chapter is devoted to carbohydrates and nucleic acids. Proteins and lipids are covered separately in Chapters 7 and 9 respectively.

Figure 1. Think of monomers like individual beads. In this picture, we have seven monomers. If we string up the beads into a bracelet, we would have one polymer.

Problem 1

Which class of biomolecule does **not** have monomers and polymers?

a) carbohydrates   b) lipids   c) proteins   d) nucleic acids

---

The term *carbohydrate* is a conjunction of *carbon* and *hydrate*. To hydrate something means to add water. Therefore, the chemical formula for carbohydrates is carbon plus water, or $(CH_2O)_N$ where N is greater than 3. Before the glaze settles over your eyes, I'll illustrate the last sentence with an example.

The most common carbohydrate is glucose with a chemical formula of $C_6H_{12}O_6$. Look carefully at that formula – six carbons, twelve hydrogens and six oxygen atoms. Isn't that just CH_2O repeated six times? Glucose, which is $C_6H_{12}O_6$, could also be written $(CH_2O)_6$.

The monomers for the carbohydrates are **monosaccharides**. The word *saccharide* has its root in the name *Saccharum*, which is the genus for sugarcane plants. A monosaccharide is a carbohydrate with between 3 and 7 carbon atoms. Glucose and fructose are the most common monosaccharides.

Problem 2

Apply the information described in the previous paragraphs to identify which of these chemical formulas could be a monosaccharide. Circle both correct answers. For each incorrect choice, provide a brief explanation for why that answer is incorrect.

$C_5H_{10}O_5$ $C_{12}H_{22}O_{11}$ $C_6H_6O_2$ $C_3H_6O_3$

On a side note......

Sugars and monosaccharides are not always the same thing. A sugar can be a monosaccharide, or a **di**saccharide (two monosaccharides linked together - the prefix "di" means two), or even a short chain of monosaccharides as long as the molecule is used to sweeten food. Although abundant, glucose is actually not that sweet and rarely consumed as a sweetener. That honor belongs to sucrose, which is table sugar. Sucrose is a disaccharide made by linking glucose and fructose together; fructose is a monosaccharide common in fruits. Molecules that are sugars have an *-ose* suffix in their name.

Monosaccharides can exist in two forms – a linear chain or a ring. Place your smartphone or tablet over the following QR code to explore the ring and linear forms of glucose, which is the most common sugar in the world.

In living systems, **sugars are almost always in rings**. For this reason, monosaccharides are almost always shown in textbooks in their ring formation rather than their linear formation. Hexagon (six-sided) and pentagon (five-sided) rings are the two most common shapes. In every ring, an oxygen atom makes up one of the points, and carbon atoms make up the remaining points. The easiest way to get familiar with monosaccharides is to draw them, as you will do the following problems.

Problem 3

Draw out the longhand structure of glucose in the white space by applying techniques covered in Chapter 4. Your final structure should have 6 carbon atoms, 12 hydrogen atoms and 6 oxygen atoms. (See Problem 10 in Chapter 4 for more help).

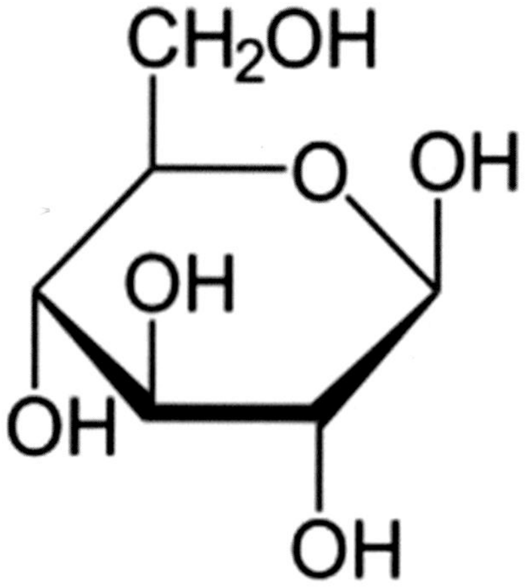

Problem 4

Draw out the longhand structure of fructose in the white space by applying techniques from Chapter 4. Your final structure should have 6 carbon atoms, 12 hydrogen atoms and 6 oxygen atoms.

You might have noticed that the chemical formula for both glucose and fructose are the same. Both molecules are $C_6H_{12}O_6$. Yet glucose exists as a six-sided ring and fructose exists as a five-sided ring, and so the sugars differ structurally. The shape of the sugar and placement of the functional groups – even something like whether the -OH group faces upwards or downwards off a carbon atom – can greatly affect how that sugar is used by an organism. We'll see this concept illustrated in a few pages.

Monosaccharides in ring formation are the **monomers**, or basic building blocks, of carbohydrates. Linking these monosaccharides together creates a **polymer**. Sucrose, which is table sugar, is created by linking glucose with fructose by a covalent bond as shown in **Figure 2**. But let's not stop there. A chain of hundreds of these monosaccharides can be built by cells to suit their needs.

To understand what those cellular needs might be, let's first talk about what role carbohydrates play in humans and pathogenic microbes – the two types of organisms you need to consider for your Human Anatomy, Human Physiology and Microbiology courses.

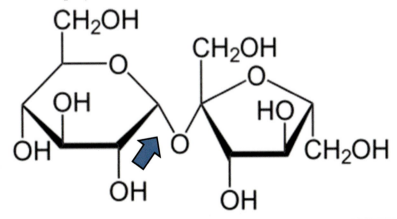

Figure 2. Sucrose molecule formed by a covalent bond (arrow) linking glucose (hexagon structure on the left) to fructose (pentagon structure on the right).

Being alive means that you have the capacity to capture and use energy. For humans and pathogenic bacteria, the source of energy is found in organic molecules ingested as food.[1] Indeed, the vast majority of the food that you eat is used as fuel, much like a car burns through gasoline. And like a car, you also require O_2 for the efficient combustion of that food and produce CO_2 as a byproduct that is expelled as waste. Humans and pathogenic bacteria consume carbohydrates, lipids, proteins and nucleic acids regularly, but *monosaccharides are the preferred fuel*. A disaccharide such as sucrose is simply broken apart into its two monosaccharides (glucose and fructose) before being offered up as cell fuel. A substantial part of our diet consists of carbohydrates that are not sugars. These are polysaccharides, a category that includes glycogen, starch and cellulose.

In living systems, **polysaccharides** (the prefix *poly-* means "many") are chains of repeating glucose monomers. Therefore, polysaccharides are the polymer form of carbohydrates. Again, each glucose molecule is a bead and the polysaccharide is a strand of those beads. **Figure 3** shows two views of a polysaccharide. In **Figure 3**, notice that not all glucose monomers are linked into a linear chain; glucose can also branch off existing chains to generate a more elaborate structure.

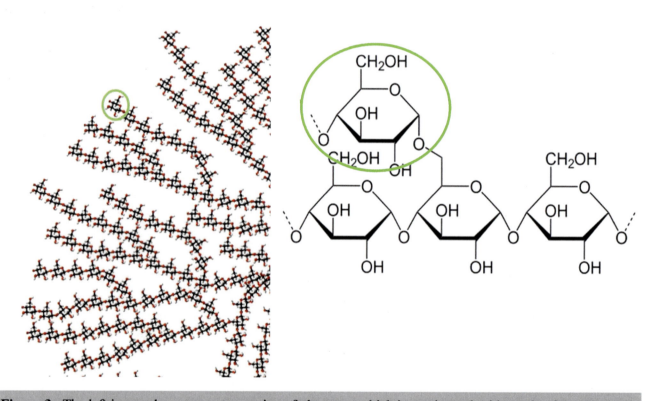

Figure 3. The left image shows a quarter section of glycogen, which is a polysaccharide made of thousands of repeating glucose units. Polysaccharides can be massive! The green circle represents in the upper left represents one glucose unit. The right image shows close up of a similar polysaccharide (starch). Note the oxygen atom that connects each glucose unit to the next one. The green circle again represents a single glucose molecule.

---

[1] If this seems like I stated the obvious, it is important to recognize that many organisms do **NOT** get their energy from consuming organic molecules. Plants, algae and other photosynthetic organisms get their energy from the sun. Some bacteria get their fuel by stripping electrons of metals and other inorganic molecules in a process known as chemosynthesis. However, humans and all pathogenic bacteria get their energy from breaking down organic molecules.

Of the three types of polysaccharides that humans *ingest*, we can only *digest* two of those – glycogen and starch. Glycogen and starch are similar polysaccharides.[2] Both are used as a "piggy bank" to store excess glucose. When animals have extra glucose, they can link up the remaining glucose into a chain for storage. This long chain of glucose is called glycogen. When said animal is running low on glucose (e.g. during fasting), the glycogen is cut back up into individual glucose monomers and then used by cells for fuel. In plants, the storage form (i.e. piggy bank form) of glucose is starch. Plant products such as grains, potatoes and rice are all very rich in starch.

When we eat animal products or softer plant products, we consume these storage polysaccharides. Our digestive system releases enzymes that easily cleave (i.e. cut) the covalent bonds holding the polysaccharide chain together. Thus, our gastrointestinal tract breaks the polysaccharides back into glucose molecules. Place your smart phone or tablet over the following QR code for a more detailed explanation.

The third type of polysaccharide that we ingest is cellulose. Cellulose is also made by plants, but it is a type of *structural* carbohydrate; it contributes to the support of the cell. Cellulose is a major component of plant cell walls and provides stiffness to the plant. When you eat a vegetable that is very rigid, like iceberg lettuce or celery, you are eating a lot of cellulose. Cellulose is also a polymer made from glucose monomers, but in cellulose, the glucose molecules are oriented differently to allow the cellulose strands to form hydrogen bonds with each other (**Figure 4**). Remember, although weak individually, hydrogen bonds in large numbers are very difficult to break. Thus, cellulose polymers assemble into one tough fiber! For example, the wood that makes the structural frame of your house is nearly 50% cellulose.

A — A — A — A — A — A — A — A — A   Orientation of glucose monomers in starch

A — ∀ — A — ∀ — A — ∀ — A — ∀ — A
A — ∀ — A — ∀ — A — ∀ — A — ∀ — A
A — ∀ — A — ∀ — A — ∀ — A — ∀ — A   Orientation of glucose monomers in cellulose
A — ∀ — A — ∀ — A — ∀ — A — ∀ — A

Figure 4. Orientation matters! Assume the letter **A** represents a glucose monomer. Orienting all of the glucose monomers in the same way creates a strand of starch as depicted in the top image. But consider flipping every other **A** upside down. This is creates a strand of cellulose (bottom image). The advantage the latter orientation is it allows the parallel strands to form many hydrogen bonds, thus creating thick ropes or fibers that lend structural support to plant cell walls. Consider this – potatoes are almost all starch, and wood is mostly cellulose. Both are plant products made from polysaccharides, but would you build a house out of potatoes? No! It wouldn't provide enough support.

---

[2] Besides the fact that starch is made by plants and glycogen is made by animals, the only structural difference is that glycogen is more branched than starch (i.e. more glucose chains branching off the main chain).

Earlier, I mentioned that a small change in chemical structure can make a big impact on the behavior of a molecule. The difference between starch and cellulose illustrates this concept in more ways than one. Not only does flipping every other glucose upside down allow for hydrogen bonds to form between parallel strands (thus creating a much stronger structure), but it turns out that the covalent bond linking each glucose in cellulose is also slightly different. The polysaccharide-digesting enzymes produced by animals only work when **every glucose is oriented in the same direction**! This would be the case with *storage* forms of glucose – starch and glycogen. However, when every other glucose is flipped upside down as is done in cellulose, the covalent bond linking two monomers is more difficult to access, and our enzymes cannot cut it. Consequently, cellulose cannot be digested by humans. Herbivores (plant-eating animals) such as cows, giraffes and even termites stock their guts with special microbes that manufacture an enzyme capable of digesting cellulose. Unfortunately, human guts are not well-populated with such microbes and therefore the energy stored in cellulose-rich foods cannot be harnessed even if consumed. Put another way - if you were lost in the woods and starving, you might be able to survive if you found berries (rich in fructose and starch), but would **not** survive off of eating bark and grass (rich in cellulose).

> *On a side note......*
>
> Dietary fibers are indigestible plant products, and **cellulose is a major dietary fiber**. Thus, when you eat food rich in cellulose (e.g. whole unprocessed grains, celery, cabbage, kale, carrots, and iceberg lettuce) you are consuming dietary fiber. Because dietary fiber is not digested, it remains in your digestive tract until it is passed as feces. This is actually very helpful as the added bulk keeps your bowel movements regular. In addition to staving off uncomfortable constipation, regular bowel movements reduce your risk of hemorrhoids and colon cancer significantly. As I am fond of saying, *a salad a day keeps the colon doctor away*.

Let's finish off our discussion of carbohydrates with two comprehensive problems. Check your answers before moving onto the next section.

Problem 5

Select the correct word to finish each sentence. Don't use an internet search engine. Try to figure out each answer by applying information presented in this chapter and previous chapters.

1. Carbohydrates are _____ molecules.
    a) polar   b) nonpolar

2. The monomer of a carbohydrate is a _____.
    a) sugar   b) monosaccharide   c) disaccharide   d) starch

3. The polymer of a carbohydrate is a _____.
    a) sugar   b) monosaccharide   c) polysaccharide   d) starch

4. Carbohydrates are _____ and therefore _____ in water.
    a) hydrophobic; not soluble   b) hydrophilic; soluble

Problem 6

For each statement, mark **true** or **false**. For any statement that is marked **false**, explain why the statement is false.

1. Plants make both starch and cellulose.

2. Starch is a polymer of glucose but cellulose is a polymer of fructose.

3. The covalent bonds of starch can be digested by humans, but the covalent bonds of cellulose and glycogen cannot be digested by humans.

4. Glycogen, starch and cellulose are all polysaccharides.

5. Sugars and monosaccharides are the same thing.

6. In living systems, monosaccharides primarily exist in linear form.

7. Carbohydrates are organic molecules.

We now turn our attention to **nucleic acids**, another type of biomolecule that adheres to the *monomer-polymer* scheme we illustrated with carbohydrates. Three important types of nucleic acids are DNA, RNA and ATP. Hopefully, these initials are tickling your memory. If not, read on.

Nucleic acids follow the monomer-polymer scheme. The *monomer* of nucleic acids is the **nucleotide**. Despite being a basic building block, the nucleotide is complex, as shown in **Figure 5**. In fact, **Figure 5** shows us that a single nucleotide can be further dissected into three sections.

Place your smartphone or tablet over the following QR code for more information about the nucleotide.

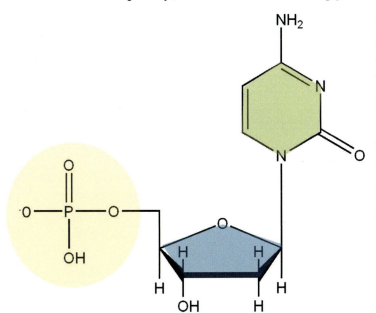

Figure 5. This entire structure represents a nucleotide from DNA. The different colors represent distinct regions. Note the blue section is actually a five-carbon sugar.

Problem 7

Evaluate the chemical structure shown in Figure 5. Are nucleic acids *hydrophilic* or *hydrophobic*? Explain your answer.

Not all nucleotides look exactly like **Figure 5**, though they all follow a similar format. For example, DNA and RNA do not use *quite* the same sugar. ATP has three phosphate groups instead of one. There are actually five different nitrogenous bases. These slight differences will be highlighted in our brief discussion of ATP, DNA and RNA (in that order).

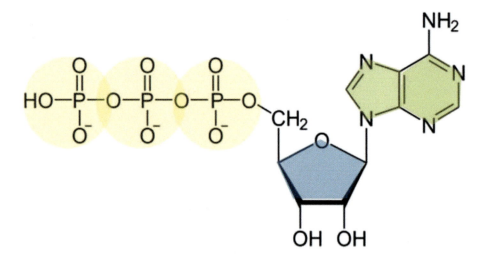

ATP stands for **A**denosine **T**ri**P**hosphate. The second word in the name is the important one. TRI... phosphate. *Tri* means three and indeed, ATP is distinguished by its three phosphate groups (**Figure 6**).

Figure 6. Adenosine TriPhosphate or ATP as it is more commonly known.

Problem 8

The left image is ATP, the right image is a nucleotide from DNA. Identify three differences – one each in the phosphate group, the 5-carbon sugar, and the nitrogenous base.

If you successfully completed Problem 8, then you noticed nitrogenous bases can be single rings or double rings. The single rings are called **pyrimidines** and the double rings are called **purines**. It's easy to confuse the names. A single ring looks like the rim of a basketball hoop, so I remember that single rings are py<u>ri</u>midines. And purines therefore, must be the double rings. Before we move onto DNA and RNA, let's look at the five nitrogenous bases we encounter in biology.

Problem 9

Use an internet search engine or textbook to find the chemical structures of the following nitrogenous bases. Draw out the chemical structure of each nitrogenous base. It is important to actually draw each structure as drawing helps you commit the information to memory. As an example, uracil has been done for you.

ADENINE   CYTOSINE   GUANINE

THYMINE   URACIL

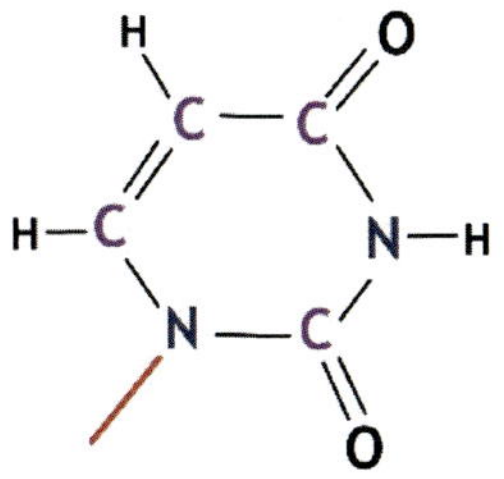

Problem 10

From your drawings in Problem 9, classify each nitrogenous base as either a pyrimidine or a purine.

PURINE: _____, _____

PYRIMIDINE: _____, _____, _____

Problem 11

Evaluate the nitrogenous bases that you drew in Problem 9. Which nitrogenous base is most similar to uracil?

  a) adenine   b) cytosine   c) guanine   d) thymine

The chemical details of a nitrogenous base are important to the function of DNA and RNA, two nucleic acid molecules that participate in cell genetics. Both DNA and RNA consist of one or two **polynucleotides** - nucleotides strung together in a strand. Because nucleotides are the *monomer* of nucleic acids, a *polynucleotide* is the polymer.

Access this QR code with your smart phone or tablet for a discussion about polynucleotide structure and the DNA double helix. Use the information conveyed in the video to tackle the next few problems. You may need to watch the video more than once to answer the next three problems.

Problem 12

Name the four nitrogenous bases that are used in DNA.

Problem 13

How many polynucleotides are in a strand of DNA? _____

Problem 14

A *base pair* consists of two nitrogenous bases facing in towards each other.

1. What type of chemical bond holds the base pair together?
    a) covalent   b) ionic   c) hydrogen

2. What nitrogenous base will pair with cytosine? _____
3. What nitrogenous base will pair with thymine? _____

As you probably recall, DNA is the genetic material of an organism. Think of the specific DNA code as the architectural blueprints for the organism. When we talk about DNA code, we are referring specifically to the pattern of nitrogenous bases. Much like a computer program is written in binary code (a specific sequence of zeros and ones), the pattern of **A**denine, **C**ytosine, **T**hymine, and **G**uanine bases is the code for an organism.

RNA is the third type of nucleic acid. Because DNA and RNA are so similar, it is convenient to highlight their few differences rather than focus on their many similarities. Here are the ways that RNA and DNA are structurally different.

1. **DNA and RNA use slightly different versions of the 5-carbon sugar.**

    RNA is short for **R**ib**o**nucleic **a**cids. Ribo stands for ribose, a 5-carbon sugar. DNA is short for **D**eoxyribonucleic acid. Note the prefix *deoxy-*, which looks suspiciously like *deoxy*genate. Indeed, DNA uses the ribose sugar with one oxygen atom removed as shown in Problem 8.

2. **DNA exists as a double strand, RNA is usually found as a single strand.**

    As described earlier, DNA in a living organism exists as two long polynucleotide strands wrapped around each other. RNA is usually a single polynucleotide strand.

3. **RNA does not use thymine, but instead uses the pyrimidine uracil.**

    One of the key differences between DNA and RNA is which bases are used. **DNA** uses A, C, G, and T. **RNA** uses A, C, G, and U. Uracil is a nitrogenous base exclusive to RNA; thymine is a nitrogenous base used only in DNA. Recall from Problem 11 that uracil and thymine are quite similar in structure.

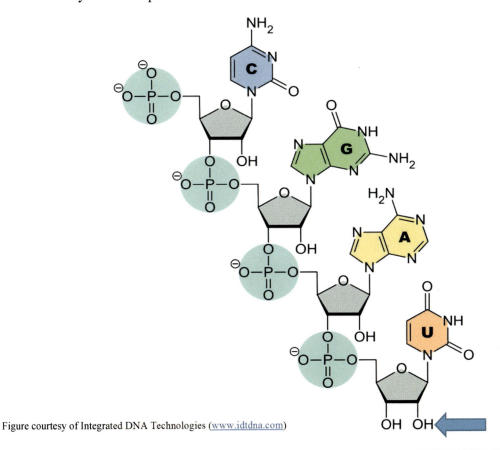

Figure courtesy of Integrated DNA Technologies (www.idtdna.com)

Figure 7. A strand of RNA illustrates the three differences between RNA and DNA. 1) The blue arrow shows the oxygen atom in the ribose sugar. Compare with a DNA nucleotide shown in Figure 5, where the oxygen atom is missing. 2) Uracil is a base instead of thymine. 3) RNA exists as a single polynucleotide.

And finally, let's give these nucleic acid molecules some framework by highlighting their functional roles in the cell. **ATP** will be explored in Chapter 10, so for now, we'll just say that ATP is the spendable form of energy in a cell. Think of **ATP** as $ that can be spent to do work.

DNA is the architectural blueprint for the organism. In a eukaryotic cell such as a human cell, almost all DNA is held inside the nucleus of a cell.[3] The DNA is organized into very long strands called **chromosomes**. A human cell contains 46 chromosomes. Within one chromosome, the DNA strand is delineated into shorter segments called **genes**. This is similar to organizing a long text into a series of paragraphs, where each paragraph represents a distinct idea. Each **gene** codes for one product – often a protein. There are tens of thousands of genes enscripted in the collective DNA code of a human cell, with anywhere from a dozen to a few thousand embedded in a single chromosome (**Figure 8**).

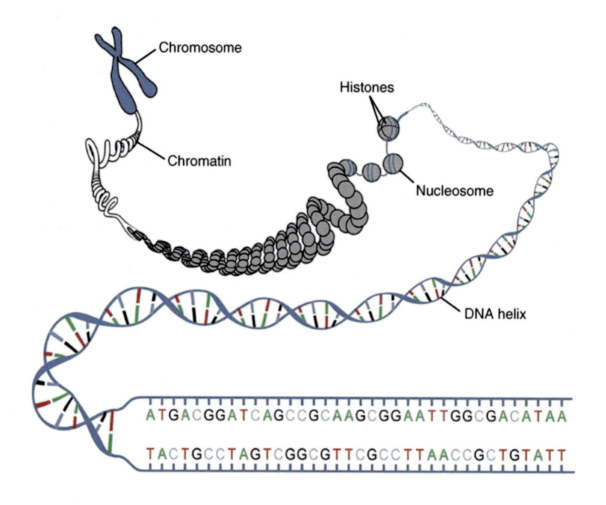

Image by OpenStax College [CC BY 3.0 (http://creativecommons.org/licenses/by/3.0)], via Wikimedia Commons

Figure 8. Very long stands of double helix DNA are organized in chromosomes. Each chromosome contains anywhere from a dozen genes to several thousand genes. Each gene is a segment of DNA that codes for some product. Often the product is a protein.

---

[3] Two organelles - mitochondria and chloroplasts - contain DNA as well.

Consider this analogy – compare all DNA in a eukaryotic cell's nucleus to a supercomputer with ten thousand software programs installed. If you accessed this supercomputer to write a paper, would you run all ten thousand software programs? No, you'd only open the word processer. The same notion applies to the DNA in a cell. A cell doesn't want to access all of the information stored in the nucleus **at the same time!** A cell should only access or **express** those genes that are needed for the cell at that moment.

For example, your liver and muscle cells contain the genes to make eyes and teeth, but those genes lay dormant in those cells. If "teeth genes" were expressed in your liver and muscle cells, you'd have teeth in your liver and muscles! That won't do!

On a side note……

This tumor, which includes teeth (red arrow) and parts of skin and hair (red circle), was taken from a woman's ovary. Sometimes, organs do express the wrong genes and the result is exactly what you'd expect - all kinds of wrong. Image by Billie Owens [CC BY-SA 3.0 via Wikimedia Commons].

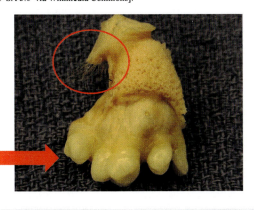

So, how does a cell express only specific genes? When one of the thousands of genes in the nucleus needs to be accessed, the cell makes a "photocopy" of that gene with an RNA strand. This is one of the major roles of RNA in the cell – to act as a mobile copy of a DNA gene that needs to be expressed. Thus, a single stranded RNA photocopy of a gene is made inside the nucleus. That RNA photocopy can then exit the nucleus and find the ribosomes – the organelles that actually translate the code into a protein. In this way, RNA is a sort of accessory molecule to human genetics, providing an avenue to give the rest of the cell a working copy of its genetic information on a *need-to-know* basis.

Problem 15

From each pair of words, circle the **larger** structure. **E.** has been done for you as an example; a cell nucleus is larger than a chromosome.

- A. MONOSACCHARIDE   NUCLEOTIDE
- B. RNA STRAND   ATP
- C. GENE   CHROMOSOME
- D. NUCLEOTIDE   DNA
- E. (NUCLEUS)   CHROMOSOME

Problem 16.   ATP stands for **A**denosine **T**ri**P**hosphate. *Based on the name*, which nitrogenous base is used in ATP? _____

Chapter 6: Answer key

1. b

2. $C_5H_{10}O_5$ are $C_3H_6O_3$ could be monosaccharides as these structures fulfill both of the criteria. Namely, that the chemical formulas are $(CH_2O)_N$ where N is a number between 3 and 7. $C_5H_{10}O_5$ is equivalent to $(CH_2O)_5$ and $C_3H_6O_3$ is equivalent to $(CH_2O)_3$. The other two choices do not fulfill this model because the ratios are off and/or the number of carbons is greater than 7.

3. [structure of monosaccharide in ring form]

4. [structure of monosaccharide in ring form]

5. 1) a   2) b   3) c   4) b

6. 1, 4, 7 are true statements; 2, 3, 5, 6 are false statements.
   #2 is false because cellulose is also a polymer of glucose.
   #3 is false because glycogen and starch can be digested by humans and only cellulose cannot be digested by humans.
   #5 is false because some sugars are disaccharides and some monosaccharides are not sugars if they do not make food sweeter.
   #6 is false because in living systems, monosaccharides usually exist in ring form.

7. **Hydrophilic.** Polar molecules and ions are both hydrophilic. The phosphate group has a negative charge and therefore is ionic. The sugar and nitrogenous base are both strongly polar because of all the polar covalent bonds.

8. ATP has three phosphate groups; the DNA nucleotide has one phosphate group.

   ATP's five-carbon sugar has a hydroxyl group off the lower right carbon; the DNA nucleotide's five carbon sugar has only a hydrogen atom off the same carbon.

   The nitrogenous bases are different with the most obvious difference being that ATP has a double ringed nitrogenous base and the DNA nucleotide has a single ring nitrogenous base.

9. (*No answer; you were instructed to look up the structures on your own*).

10. Purine (double rings) – adenine, guanine; Pyrimidine (single rings) – cytosine, thymine, uracil

11. d

12. adenine, cytosine, guanine, and thymine

13. 2

14. A. hydrogen   B. guanine   C. adenine

15. MONOSACCHARIDE   (NUCLEOTIDE)
    (RNA STRAND)   ATP
    GENE   (CHROMOSOME)
    NUCLEOTIDE   (DNA)

Explanation:

Nucleotides and monosaccharides are both monomers, but nucleotides are larger for they contain a monosaccharide plus two other parts (the nitrogenous base and phosphate group). An RNA strand is made up of multiple nucleotides and therefore is larger than ATP, which is a single nucleotide. A chromosome includes many genes, and therefore is larger than a single gene. DNA is made up of thousands of nucleotides, and therefore is larger than a nucleotide.

16. The "**A**" in ATP stands for **A**denosine, which indicates the nitrogenous base *adenine*.

Chapter 7: Proteins

Proteins are like puzzles. One misshaped piece and the whole picture falls apart.

This chapter continues of our coverage of biomolecules with proteins. Proteins are an important class of biomolecules and deserve considerable attention.

Like carbohydrates and nucleic acids, proteins follow the *monomer-polymer* scheme outlined in Chapter 6. The monomer for a protein is an **amino acid,** a molecule you've seen previously in Chapters 3 and 4. Notably, there are a whopping twenty amino acids that are universal! Before cringing at the thought of learning 20 different molecules, let me explain what the endgame is.

Your endgame is to understand how a protein folds into its globular 3D shape.

Why? For one, this knowledge can be applied to **infectious microbes**. All infectious agents, including bacteria and viruses, must have correctly folded proteins to function. Without functioning proteins, these microbes cannot cause infections. Therefore, if we can disrupt the folding of these proteins, we can "kill"[1] these microbes and prevent potential infections. This is particularly important within health care settings where microbes might be lurking on surfaces, waiting for a chance to invade the human body (e.g. in a wound, puncture or surgical incision). Thus, many of the antimicrobial agents applied routinely in health care settings – alcohol-based hand sanitizer, iodine and hydrogen peroxide – work by unravelling, or **denaturing**, the proteins of microbes.

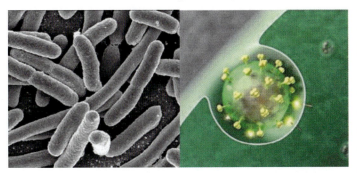

Figure 1. Left - Electron micrograph of *Escherichia coli* (*E. coli*). Right – HIV virus emerging from a human cell. HIV Image by J. Robert Trujillo released to Wikimedia Commons.

To understand how these antimicrobial agents work (which will be an important topic in your Microbiology course), you need to first address this question. **How do proteins fold into their 3D shape?** The answer starts with the chemical composition of the amino acids.

All twenty amino acids share a common structure; this is sometimes referred to as the N-C-C backbone which is shown in black ink in **Figure 2**. The part where the amino acids differ is denoted by the purple **R** (R is for residue). The **R** is a placeholder for some chemical structure, and each of the twenty amino acids has a different residue (i.e. chemical structure). If you're already lost, don't worry. Some examples and problems will help.

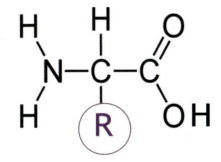

Figure 2. An amino acid. R is a place-holder for a residue (i.e. chemical structure).

---

[1] I use the word "kill" with quotes because something must first be alive in order to kill it. Bacteria can be killed; viruses (which are non-living but infectious) are inactivated. However, saying that you've inactivated the virus just isn't as satisfying as saying "I've killed the virus! Mwahahaha".

Problem 1

This amino acid is glycine. Copy the structure of glycine in the white space. Then, circle all atoms that are part of the N-C-C backbone. Put a box around the *residue*.

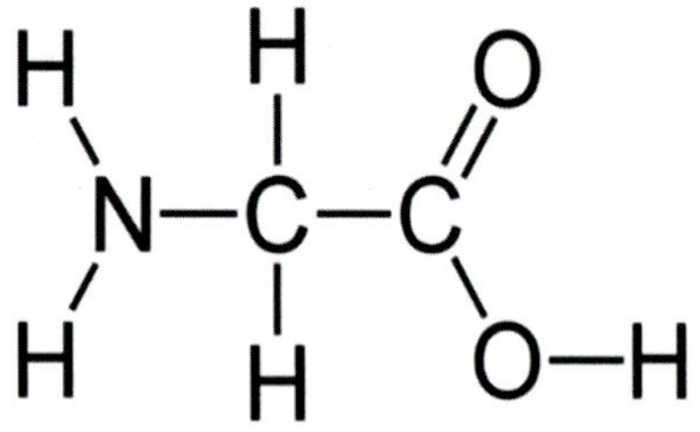

Problem 2

Here are four amino acids. In each amino acid, circle all atoms that are part of the N-C-C backbone and box the *residues*.

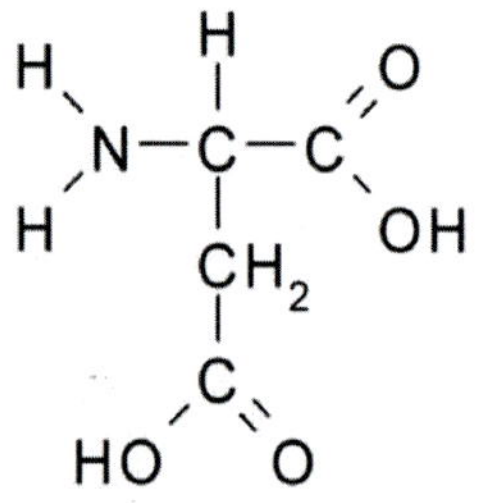

Aspartic acid

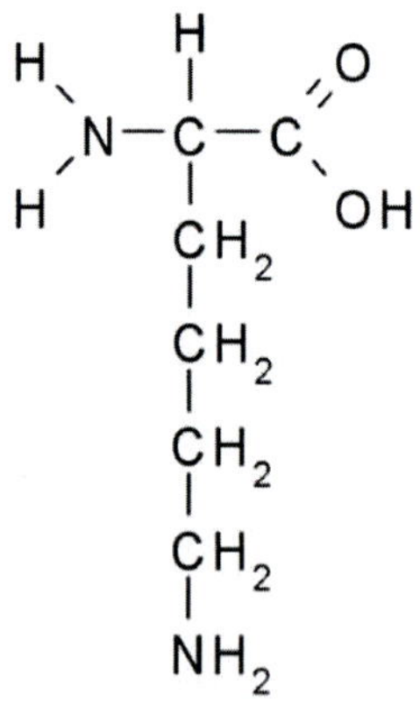

Lysine

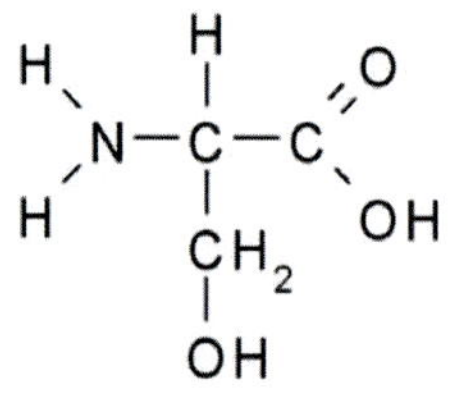

Serine

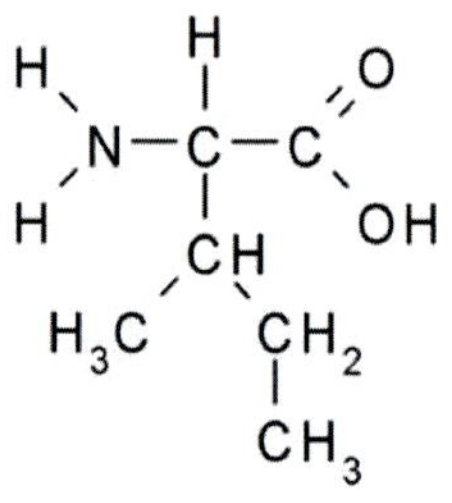

Isoleucine

All amino acid images are licensed under CC BY-SA 3.0 via Wikimedia Commons

Problem 3

Amino acids can also be drawn in shorthand. Name the two amino acids shown here. Each shorthand structure is one of the four amino acids presented in Problem 2.

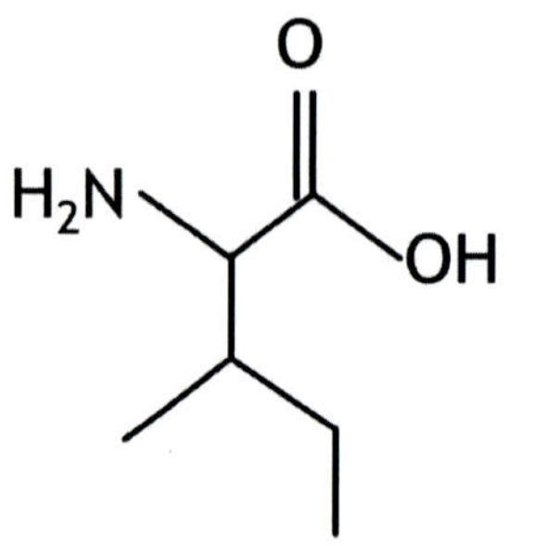

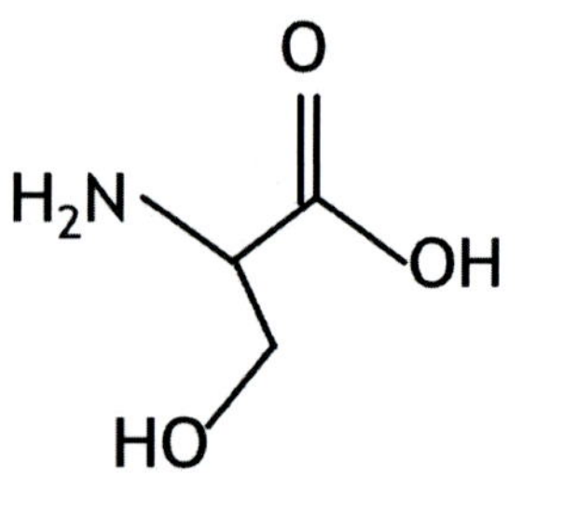

_____ _____

The twenty amino acids are sorted into one of these categories **based on the chemical structure of their residue**. The categories are:

1. POLAR
2. NONPOLAR
3. ELECTRICALLY CHARGED
   a) Acidic
   b) Basic

This tutorial will explain how to classify amino acids based on their residue. Please place a smartphone or tablet over the following QR code for the explanation.

Now it's your turn. Apply the information presented in the video to classify each of the following amino acids as *polar*, *nonpolar* or *electrically charged*. Remember to apply your knowledge. As tempting as it may be, searching for answers on the internet does little to help you learn.

Problem 4

Classify each amino acid as **polar**, **nonpolar** or **electrically charged**. If the amino acid is electrically charged, further identify the amino acid is **acidic** or **basic**.

1. [structure with CH₂–C(=O)–O⁻ side chain]
2. [structure with CH₂–OH side chain]
3. [structure with CH(CH₃)–CH₂–CH₃ side chain]
4. [structure with CH₂–CH₂–CH₂–NH–C(NH₂)=NH₂⁺ side chain]
5. [structure with CH(CH₃)(CH₃) side chain]
6. [structure with CH₂–C(=O)–NH₂ side chain]

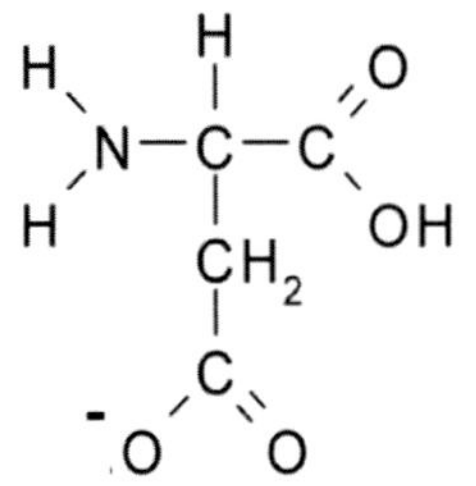

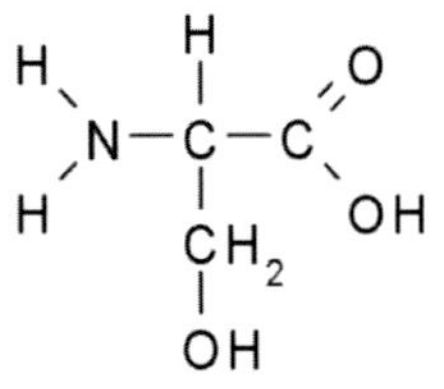

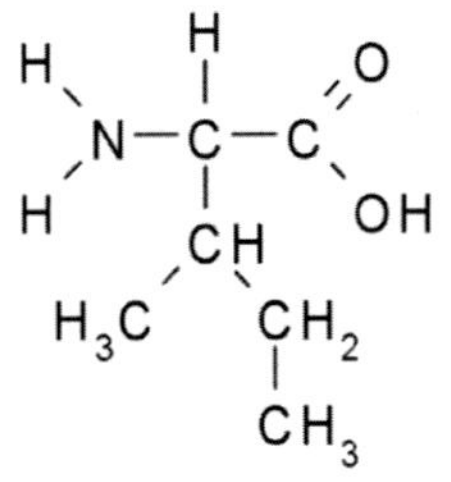

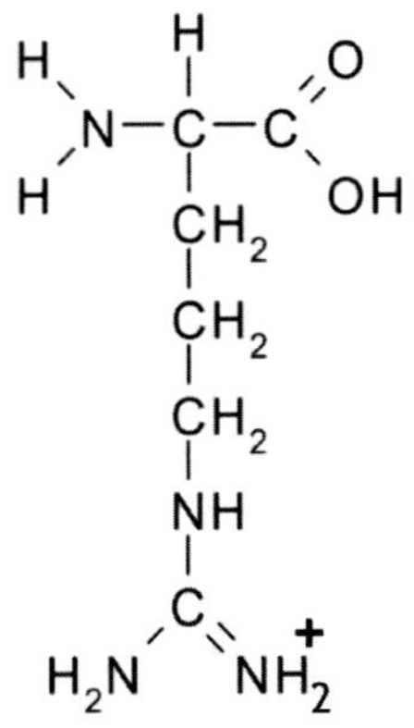

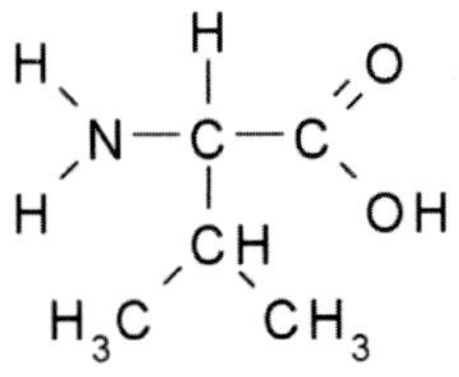

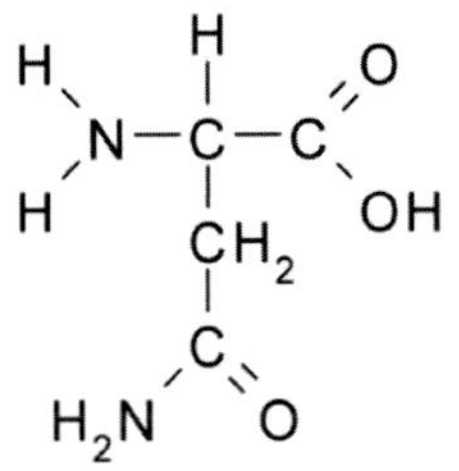

All amino acid images are licensed under CC BY-SA 3.0 via Wikimedia Commons

1. _____   2. _____   3. _____
4. _____   5. _____   6. _____

79

Amino acids are the monomer of proteins, and there is quite a variety of them. The polymer emerges when amino acids are linked together by covalent bonds. A short chain of amino acids is a **peptide**. A long chain (more than 50) of amino acids is called a **_poly_peptide**. Recall that the prefix "*poly*" means *many*; in this case, a polypeptide has *many* amino acids. The peptide or polypeptide is the polymer form of the protein. Place your smartphone or tablet over the QR code on the left to watch an animation on how to build a peptide from three amino acids. Then try Problem 5.

Problem 5

Answer the following questions concerning this peptide of three amino acids: *Leucine – Alanine – Phenylalanine.*

$$H_2N-CH(CH_2CH(CH_3)_2)-CO-NH-CH(CH_3)-CO-NH-CH(CH_2C_6H_5)-COOH$$

A. Circle the two **peptide bonds**.

B. Put a box around each residue.

C. These amino acids are _____.

    a) polar   b) nonpolar   c) electrically charged; basic
    d) electrically charged; acidic

D. The residues of these amino acids are _____.

    a) hydrophilic   b) hydrophobic

A peptide or polypeptide structure does not stay linear; linear chain of amino acids would have little use to an organism. **The value of the polypeptide manifests when the polypeptide coils and folds into a specific 3D form**. Only then does the molecule have utility as a **protein**. Proteins are one or more folded polypeptides that have a specific function in the cell. Although some proteins are purely structural (e.g. cytoskeleton – the fibers that provide shape to an animal cell), most proteins are workhorses. Enzymes are proteins that assist chemical reactions, and transporters are proteins that move molecules across a membrane.

Place your smart phone or tablet over the following QR code for a detailed introduction to protein folding.

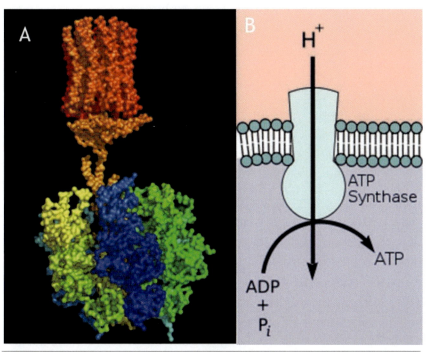

Figure 3. ATP synthase – the protein that synthesizes ATP in a cell - is a huge protein consisting of thousands of amino acids on at least eight different polypeptides. **A.** The specific 3D globular shape, each polypeptide is a different color. **B.** The cartoon version of ATP synthase typically shown in textbooks.

Figure 3 shows two views of ATP synthase – a very large protein that consists of at least eight polypeptides (each a different color in **Figure 3A**). **Figure 3A** shows the fine details the functional protein's configuration. **Figure 3B** shows the cartoonish version of ATP synthase as it is usually presented in a general biology textbook. Note that the nuances of the 3D shape (i.e. each curve and detail) are lost in the cartoon drawing of **Figure 3B**.

The anatomy, or 3D shape, of the protein matters immensely. Change the shape a little bit, and the protein function is altered or lost. This point bears repeating. Think of it like changing your house key just "a little" and then trying to open your front door. What may seem like a trivial, insignificant change to your key might be the difference between the key working and being locked out of your house. The analogy holds true for most proteins.

The structure of hemoglobin, the large protein holds O_2 in red blood cells, will illustrate this point. Place your smartphone or tablet over the following QR code to see how small changes in hemoglobin structure can alter the color of your blood.

Problem 6

A functional protein is _____.

   a) the same thing as a polypeptide
   b) one or more polypeptides folded into any shape
   c) one or more polypeptides folded into a specific shape

A functional protein must have correctly folded polypeptide(s). How a (poly)peptide is folded is therefore a subject worth learning. As explained in a previous tutorial, (poly)peptide folding is not random; it is the outcome of chemical interactions between amino acids within the chain. The following lists the types of chemical interactions that we've learned about in Chapter 2:

1. covalent bonds
2. ionic bonds
3. hydrogen bonds

All of those chemical bonds play a role in protein folding and configuration. There is a fourth type of chemical interaction that plays a significant role protein folding and that is the **hydrophobic interaction**.

The hydrophobic interaction is the attraction of nonpolar molecules (or nonpolar parts of molecules) towards each other in the presence of water. Remember, nonpolar molecules are hydrophobic. In the presence of water, these nonpolar molecules are compelled to stay together – a *safety in numbers* approach in sea of enemies (assuming water is the "enemy"). For example, put a few drops of oil (a nonpolar and hydrophobic molecule) in a pot of water and stir. The stirring action may separate the oil into smaller droplets, but as soon as you stop stirring, the oil droplets will rejoin into a single glob. These are hydrophobic interactions at work. Another example is oil and vinegar dressing in a bottle. Leave the bottle alone and the oil will cleanly separate from the vinegar (which is mostly water plus a little acetic acid). The oil molecules aggregate together out of fear of water.

Problem 7

What part of the polypeptide will participate in hydrophobic interactions during protein folding?

   a) nonpolar residues only
   b) polar residues only
   c) electrically charged residues only
   d) the N-C-C backbone only
   e) Nonpolar and electrically charged residues will both participate in hydrophobic interactions.

Problem 8

What part of the polypeptide may participate in ionic bonds during protein folding?

a) nonpolar residues only
b) polar residues only
c) electrically charged residues only
d) the N-C-C backbone only
e) Nonpolar and electrically charged residues will form ionic bonds with each other.

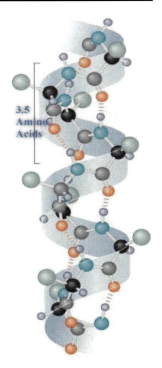

Ensure your answers to Problems 7 and 8 are correct before moving on.

In Chapter 2, we learned that hydrogen atoms in polar covalent bonds can participate in a second type of chemical interaction – a hydrogen bond. So what parts of a polypeptide might form hydrogen bonds? **Polar residues** are good candidates, but so is the N-C-C backbone, which has a δ^+ (partially positive) hydrogen atom attached to each nitrogen. In fact, the N-C-C backbone will often coil around in a helical formation to create a spiral (**Figure 4**). That spiral shape (called an **alpha helix**) is held together by hydrogen bonds.

Figure 4. The N-C-C backbone of the peptide chain coils around itself to form a helical structure (alpha helix). The hydrogen bonds are between a δ^+ hydrogen (small purple sphere) and the δ^- oxygen atom.

Problem 9

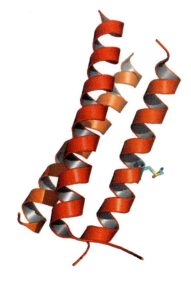

This image shows three alpha helices. What chemical interactions holds the spiral pattern of an alpha helix in place?

a) covalent bonds between residues
b) hydrogen bonds between *N-C-C backbone* atoms
c) hydrophobic interactions between nonpolar residues
d) ionic bonds between acidic and basic functional groups

Image credit. "Alphabody1" by Paul RHJ Timmers - PDB ID: 4OG9 Desmet, Johan et al. (5 Feb 2014). "Structural basis of IL-23 antagonism by an Alphabody protein scaffold". Nature communications 5: 5237. doi:10.1038/ncomms6237. Licensed under CC BY 4.0 via Wikimedia Commons - http://commons.wikimedia.org/wiki/File:Alphabody1.png#/media/File:Alphabody1.png

The use of covalent bonds in maintaining protein structure is limited to proteins where at least two cysteine amino acids are found. Cysteine is unusual in that it contains a sulfur atom in its residue (**Figure 5**). The sulfur atoms on two different cysteine residues can join together to form a covalent bond. This strong linkage has a special name – it is called a **disulfide bridge** (**Figure 6**). Because covalent bonds are very strong, disulfide bridges are not easily broken. Thus, disulfide bridges are useful in keeping the protein folded, especially when proteins are secreted outside of cells[2].

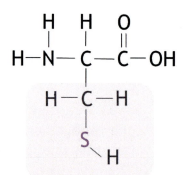

Figure 5. The amino acid cysteine.

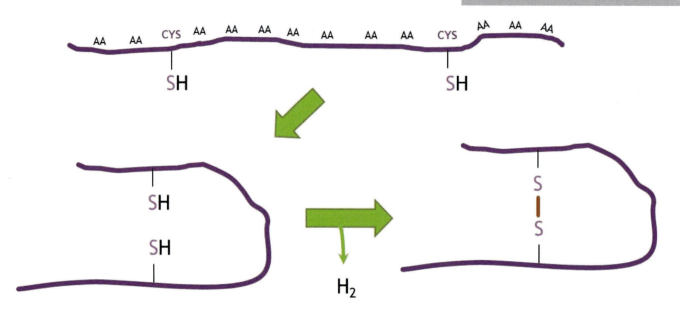

Figure 6. Top image – a peptide with 14 amino acids (AA), two of which are cysteine (CYS). Only the –**SH** part of the cysteine residues are shown. The rest of the peptide details (backbone and residues) are not shown. Left image – the peptide wraps around so the two –SH groups line up. Right image – an enzyme removes two hydrogen atoms to create a covalent bond between the two sulfur atoms. This covalent bond is the **disulfide bridge**. *Di-* means two (as in two sulfur atoms), *sulfide* means the sulfur atoms, and *bridge* refers to the covalent bond connecting the two sulfur atoms. The peptide now has a hairpin turn that is held in place by this disulfide bridge.

The next video provides a visual reinforcement of chemical interactions involved in protein structure. In this video, we explore the structure of insulin, a peptide hormone secreted by the pancreas. Place your smartphone or tablet over the QR code on the left. After watching the video, I encourage you to explore protein structure yourself by going to http://www.rcsb.org/.

---

[2] Secretory proteins, or proteins that are secreted outside of cells (rather than proteins that function within cells), might be exposed to different environments. Changes in the environment (such as additional heat or changes in pH) can dismantle weaker bonds such as hydrogen bonds or hydrophobic interactions, but not usually disulfide bridges. Thus, many secretory proteins incorporate disulfide bridges to maintain the correct folding and function of the protein outside the cell.

Problem 10

A disulfide bridge is a type of _____ bond between _____.

  a) hydrogen; sulfur atoms
  b) covalent; sulfur atoms
  c) hydrogen; hydrogen atoms
  d) covalent; hydrogen atoms

Problem 11 *Challenge!*

Match each amino acid from the word bank below with the type of chemical interaction the residue is likely to participate in. Each amino acid will be used only once. *You will need an internet search engine or biology textbook to evaluate the structure of each amino acid. As an example, two amino acids have been correctly matched for you.*

Disulfide bridges	Hydrogen bonds	Ionic bonds	Hydrophobic interactions
Cysteine			Valine

Leucine   Serine   Glutamine   Glutamic acid (Glutamate)

Lysine   Phenylalanine   Arginine

Think of a protein like an orchestra's rendition of a difficult musical piece; each chemical interaction is a musician with their instrument. Each instrument within an orchestra contributes in a specified way to the musical piece. If one instrument is off key, the music sounds different, and usually not in a pleasant way. If several instruments are off key, miss their timing, or don't play, the musical piece might become unrecognizable.

In the same way, each chemical interaction contributes to the final protein structure in a specified way. Since protein structure is the collective result of tens to hundreds of chemical interactions, it follows that changing (or eliminating) even just one of the prescribed chemical interactions can impact the protein structure. Did you get all that? This is important. Even slight changes to protein structure can significantly impact the functionality of the protein. The altered protein may function differently, less efficiently, or just not at all (think of the house key analogy).

Our last topic focuses on two causes of incorrect protein structure: DNA mutations and forces that cause denaturation.

The amino acid sequence of a (polypeptide) is the primary factor in determining what the final 3D structure of the protein will be. For example, consider a hypothetical polypeptide with 100 amino acids. Cysteine is coded for at positions 30 and 60 along the 100-amino acid chain, allowing for a disulfide bridge to connect amino acids 30 and 60 together. What happens if a DNA mutation occurs that codes for a different amino acid at position 30? In that case, the polypeptide cannot make a disulfide bridge, and the protein shape would be significantly different (**Figure 7**).

As **Figure 7** demonstrates, DNA mutations can have significant impacts on protein structure when the sequence of the amino acids is changed even at one spot. Many genetic disorders are the result of a DNA mutation that causes a single amino acid substitution in just one protein. Cancer can result when certain proteins that regulate (*i.e.* stop) cellular division are mutated in a similar fashion; such mutations cause these regulatory proteins to lose function. The loss of cellular division regulation leads to cells that divide out of control, and this creates an unwanted mass known as a tumor.

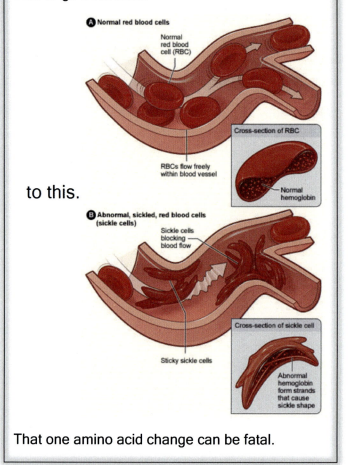

On a side note......

Serious genetic disorders can arise from proteins that have a single amino acid substitution. One famous example is sickle cell anemia. Individuals with sickle cell anemia build their hemoglobin with valine (nonpolar) instead of glutamic acid, in one of the hemoglobin subunits. That seemingly trivial change causes hemoglobin in red blood cells to go from this....

to this.

That one amino acid change can be fatal.

Figure 7. A - The shape of a hypothetical protein with a disulfide bridge (yellow) between two cysteine amino acids at the 30th and 60th positions in the polypeptide. **B** - The same protein with a different amino acid at position 30. The protein shape is changed drastically because there is no disulfide bridge without a cysteine at position 30.

In a Microbiology course, you'll learn about various antimicrobial agents. These are chemical or physical methods that are applied with the purpose of killing microbes. Many of these commonly applied agents –

hydrogen peroxide, alcohol, heat, strong acids and bases, and iodine - work by **denaturing** (or unraveling) the microbe's proteins. Denaturation results when some of the existing chemical interactions are disrupted.

For example, heat breaks weaker chemical bonds such as hydrogen bonds and hydrophobic interactions. Applying heat to contaminating bacteria will cause at least some hydrogen bonds and hydrophobic interactions to dissociate. This results in partial unraveling of the microbial proteins. Protein denaturation leads to death for the microbes, as their enzymes, transporters and other proteins no longer work. For humans, our proteins function best near our body temperature of 98.6°F. At this temperature, our proteins are correctly folded. Death by hyperthermia can result when our bodies get too hot. Our proteins begin to denature as weak chemical interactions dissociate, resulting in loss of protein function and cell death.

Problem 12

As described in the previous "*On a side note....*", sickle cell anemia is a genetic disorder caused by one amino acid substitition. At one spot along a polypeptide, glutamic acid (normal) is replaced by valine (sickle cell). Answer the following questions about these amino acids.

1. Look up the chemical structure of glutamic acid. What type of amino acid is glutamic acid?
    a) polar
    b) nonpolar
    c) electrically charged; basic
    d) electrically charged; acidic

2. What type of chemical interaction is glutamic acid likely to participate in?
    a) hydrogen bond   c) disulfide bridge
    b) ionic bond   d) hydrophobic interaction

3. Is glutamic acid hydro*phobic* or hydro*philic*?
    a) It is hydrophobic and therefore will want to move internally into the protein.
    b) It is hydrophilic and therefore will be comfortable near the surface of the protien.

4. Valine is a nonpolar amino acid. Explain why substituting valine for glutamic acid can change the shape of hemoglobin.

Chapter 7: Answer key

1.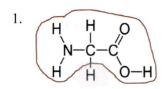

2.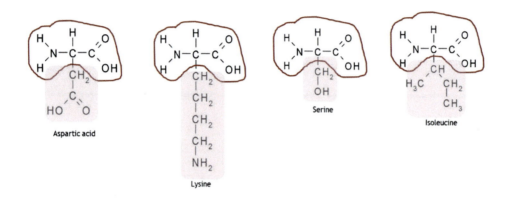

3. Left image - isoleucine; right image - serine.

4. 1) electrically charged; acidic
   2) polar
   3) nonpolar
   4) electrically charged; basic
   5) nonpolar
   6) polar

   For a detailed solution to Problem 4, place your smartphone or tablet over this QR Code.

5. A/B.

   ![structure]

   C. b) nonpolar
   D. b) hydrophobic

   Explanation: The residues (boxed) are all nonpolar because they consist of only carbon and hydrogen atoms. This means these covalent bonds are all nonpolar covalent bonds. Since they are nonpolar, the residues are also hydrophobic.

6. c

7. a

   Explanation: In order to participate in a hydrophobic interaction, the molecule must have a "fear" of water. Only nonpolar residues recoil within the presence of water. All other residues (polar, electrically charged) are hydrophilic.

8. c

   Explanation: Ionic bonds only form between two oppositely charged ions. Only the electrically charged residues satisfy that criterion. Acidic groups are negatively charged, and basic groups are positively charged.

9. b

10. b

11. Disulfide bridges (cysteine)
    Hydrogen bonds (serine, glutamine)
    Ionic bonds (glutamic acid, arginine, lysine)
    Hydrophobic interactions (valine, leucine, phenylalanine)

    Explanation: Cysteine amino acids are the only amino acids to participate in disulfide bridges. Any amino acid that is polar (but not charged) can have its residue participate in a hydrogen bond. Serine and glutamine are both polar, but not charged. Any amino acid with a net charge (ion) on its residue can participate in ionic bonds. This includes the negatively charged acids, such as glutamic acid (or glutamate), and the positively charged bases – arginine and lysine. Any amino acid that is mostly nonpolar can participate in hydrophobic interactions. Valine, leucine and phenylalanine are all nonpolar amino acids.

12. 1) d   2) b   3) b
    4) Glutamic acid is negatively charged and therefore likely to found near the surface, possibly participating in an ionic bond with some positively charged residue. When valine is substituted for glutamic acid, valine behaves very differently as a nonpolar amino acid. Valine's residue will want nothing to do with the surface or another + charged residue, and instead may pull towards the center of the protein to engage in hydrophobic interactions. This pull into the center of the protein will change the shape of hemoglobin as the folding pattern is now altered.

Chapter 8: Cells and histology

You are not yourself today – you are 90% bacteria.

A cell is the smallest unit of life. Some organisms, such as bacteria, are only a single cell. An adult human consists of trillions of cells. Cells are marvelous and complex little entities – capable of dividing to produce two cells, capturing and using energy, responding to the environment outside of the cell, and even working in concert with other cells to achieve some function.

Figure 1. Cell or fried egg?

This chapter starts with those wonderful cells and works up to **histology**, which is studying tissues under a microscope.

When I ask students to draw a cell, I often get a picture of a fried egg (**Figure 1**). Unfortunately, this "fried egg" depiction belies the dynamic complexity of these microscopic entities. For a better idea of just how much goes on in our cells, check out the wonderful "The Inner Life of the Cell" animation created by XVIVO Scientific Animation for Harvard University. You can access the video with the QR code on the right, or directly through the XVIVO website.

http://www.xvivo.net/animation/the-inner-life-of-the-cell/

The video is well worth the three minutes of your time; it follows what happens *inside* a white blood cell as the white blood cell leaves the blood.[1] After watching the video, hopefully you can appreciate that cells are much more than the boring fried eggs a student sees under a microscope.

The basic formatting of cells starts with these three components – a region containing DNA (organized as one or more chromosomes), a boundary in the form of the **plasma** (or **cell**) **membrane**, and everything in between (**cytoplasm**) as shown in **Figure 2**.

Cells come in two forms: prokaryotic and eukaryotic. **Prokaryotic cells** are the cell types found in bacteria; they are the simpler of the two cell types. A single prokaryotic cell is the entire organism, so when you look at a bacterial cell, you are looking at one organism. Prokaryotic cells are also much smaller than eukaryotic cells. For example, a human liver cell (a eukaryotic cell) is about 10,000 times larger than an *E. coli* cell by volume! Prokaryotic cells are also non-compartmentalized, which is in stark contrast to eukaryotic cells.

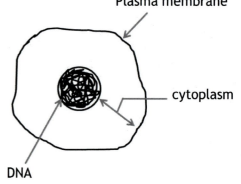

Figure 2. Cells have a region of DNA, a plasma membrane and cytoplasm.

---

[1] The intended take away from the video is that your cells are incredibly complex and dynamic. If you couldn't figure out what was happening in the video, don't worry! You can type in "Narrated version XVIVO Inner Life of Cell" into YouTube search engine to find the narrated and longer version if you so desire.

Problem 1

This is a stained image of three human (eukaryotic) cells from my gumline (before brushing my teeth). The green line indicates the boundary of one of my gum cells. **B** represents a cluster of bacteria[2]. Answer the following questions.

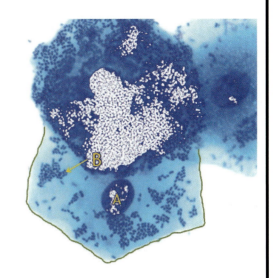

1. What structure is represented by **A**?

2. What type of cell is a bacterium?
    a) prokaryotic   b) eukaryotic

3. What is larger, a eukaryotic nucleus or a prokaryotic cell?

4. How many bacteria are visible on my cell marked by the letter **A**?
    a) 1   b) 8   c) between 12 and 15   d) over 100

Larger organisms are built from more complex **eukaryotic cells**. The defining feature of a eukaryotic cell is an organelle called a **nucleus**. However, it is the eukaryotic cell's ability to compartmentalize that is truly the hallmark of the eukaryote. It is useful to think of a eukaryotic cell like a large business. A large business will divide up its tasks among various departments to increase efficiency (e.g. IT, marketing, human resources, sales). Likewise, a eukaryotic cell delegates its tasks to different organelles.

This compartmentalization allows for specialization.

On a side note......

Humans are born sterile, but microbes colonize the skin and gastrointestinal tract within a week of birth. The colonization establishes a **microflora**, a community of microbes that inhabit an area of the human body. If you were to count the number of bacteria on your persons right now, there would be almost 10x the number of bacteria compared to the number of your own eukaryotic cells. (Remember, bacteria are 1/10,000 the size of your own cells). When someone remarks, *"you know, you're really not yourself today."* You can now retort, *"I'm not myself everyday - 90% of my body is bacteria cells!"* It's a joke only a microbiologist can appreciate. Nerd out peoples!

---

[2] Before you get too horrified at the prevalence of microbes in my mouth, I will assure you that unless you've used an antiseptic mouthwash or brushed your teeth in the last few minutes, your mouth has a similar microbial presence. ☺

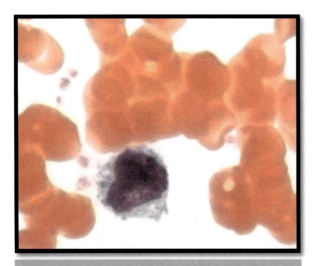

Figure 3. The red masses are red blood cells and the large white and purple mass is a monocyte (a type of white blood cell). The differences in their cell structure reflect their different roles in the body.

The ability to create different cell types is required for complex and large organisms (e.g. humans) to function. By some counts, there are over three hundred different types of human cells (**Figure 3**).

We first review the various components of a eukaryotic cell. Then, we'll apply that knowledge to explore specialization of human cells. This **structure-function relationship** is a very important principle in your studies of human anatomy and physiology.

Place your smart phone or tablet over the following QR code for a fifteen minute tour of a human cell. Then answer the following questions.

Problem 2

On this drawing of a eukaryotic cell, label each organelle or structure.

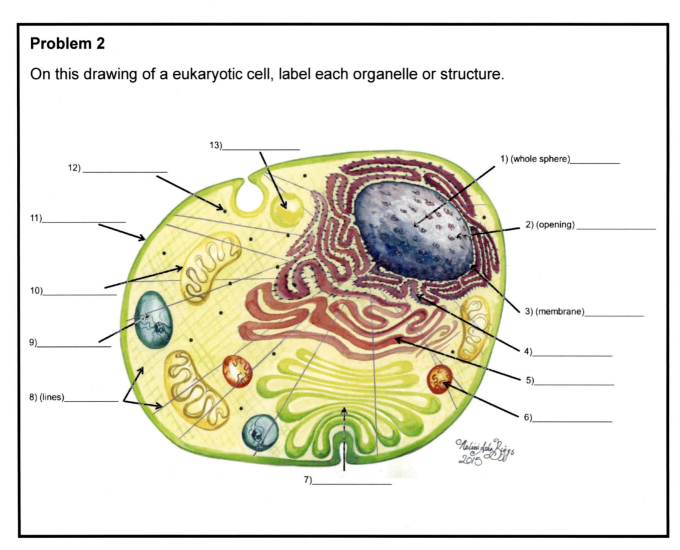

1) (whole sphere)_____
2) (opening)_____
3) (membrane)_____
4)_____
5)_____
6)_____
7)_____
8) (lines)_____
9)_____
10)_____
11)_____
12)_____
13)_____

Problem 3

Name the other five organelles that participate in the **endomembrane system**.

1) ____vesicles_____   4) _____

2) _____   5) _____

3) _____   6) _____

Problem 4

Name the two compartments within a mitochondrion.

Problem 5

Match the following cellular organelles or structures with their function. As an example, one has been done for you.

Organelle	Function
Smooth endoplasmic reticulum	synthesizes ATP from energy stored in biomolecules
Nuclear envelope	synthesizes lipids
Mitochondria	transports stuff among the endomembrane system
Ribosomes	internal structure and highway system for vesicles
Rough endoplasmic reticulum	folds polypeptides into proteins
Cytoskeleton	"post office"; ships out packages of molecules
Vesicles	breaks down, recycles particles and old molecules
Golgi body	neutralize hydrogen peroxide, break down big fats
Lysosome	allows mRNA to exit nucleus
Peroxisome	membrane that surrounds chromosomes
Nuclear pore	reads mRNA instruction, builds (poly)peptide

(Mitochondria is matched to "synthesizes ATP from energy stored in biomolecules")

Let's turn our focus to the *structure-function* relationship as it applies to cells. The principle is simple. The structure of something reflects its function. For example, the *function* of a tire is to rotate on the ground. Therefore the *structure* of a tire is round, rather than a less useful form like a square.

Teeth are excellent for demonstrating this principle. **Figure 4** shows the jaws of two different sharks with different teeth structure. The differences in structure reflect differences in function. Port Jackson sharks eat crunchy invertebrates like sea urchins; their teeth are well-suited for grinding up crunchy exoskeletons to eat the soft fleshy bodies within.

Mako sharks eat fast moving fish. And if you've ever handled a live fish, you know they are slippery. Thus, mako sharks are equipped with thin, sharp spears for teeth – perfect for gripping that fast, slippery fish in open water.

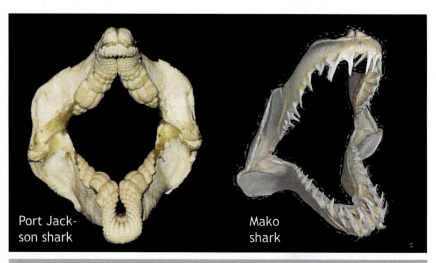

Figure 4. Jaws of a Port Jackson shark and mako shark demonstrating the structure-function relationship.

Mako shark jaw photo attributed to "Isurus oxyrinchus Machoire" by Didier Descouens - Own work. Port Jackson shark jaw photo attributed to "Port Jackson Shark Jaw 1" by Jason7825. Both images licensed under CC BY-SA 3.0 via Wikimedia Commons.

Within a human body, each cell specializes its structure by altering the amount or size of various organelles. Here are some examples of how specialized structures fit the cell's function.

Mature red blood cells (erythrocytes) have one basic function – to transport gases (mostly oxygen[3]). Hemoglobin is a large protein complex responsible for O_2 transport, and each hemoglobin molecule can hold up to four O_2 molecules. Thus, if you want a erythrocyte to transport 1000 O_2 molecules, it needs to hold at least 250 hemoglobin molecules. If a red blood cell needs to transport 100,000 O_2 molecules, then it needs at least 25,000 hemoglobin molecules. Get the idea? The more hemoglobin packed into a red blood cell, the more oxygen it can transport. Ideally, a red blood cell would just pack millions and millions of hemoglobin into its cytoplasm. Sounds good on paper, but the problem is the red blood cell would be too large. For reasons that you'll learn about in Human Physiology, red blood cells must stay small – even compared to other human cells – to function properly. What a conundrum! The red blood cell must stay small while carrying as much bulky hemoglobin as possible.

The solution? A mature red blood cell disposes with some of the larger organelles, including the nucleus (!!) and mitochondria, to make room for more hemoglobin.

Figure 5A shows the typical biconcave shape of a erythrocyte, with the middle of the cell looking quite deflated (*well*, it did lose its nucleus). **Figure 5B** shows how much space the nucleus and mitochondria occupy in a typical cell; the space-saving advantage of eliminating them is evident. (See also **Figure 3** to confirm the relatively large volume of a nucleus [purple blob] in white blood cell).

---

[3] Although the primary job of a erythrocyte is to transport oxygen from the lungs to the tissue, red blood cells also participate in the transport of carbon dioxide from the tissues back to the lungs.

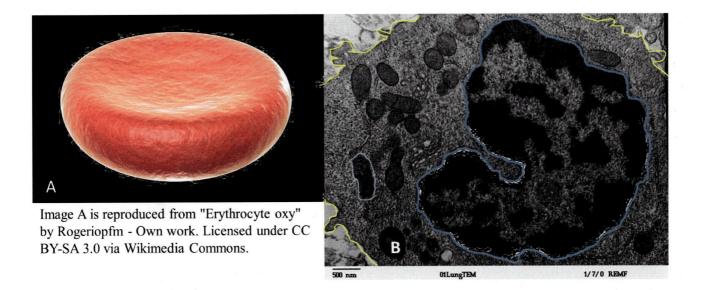

Figure 5. **A.** Surface contours of a red blood cell, demonstrating the concave surface area made possible by the lack of nucleus. **B.** Electron micrograph of a white blood cell (cross section). Plasma membrane is outlined in yellow; nuclear envelope is in blue; one (of many) mitochondrion in violet. Red blood cells free up a lot of space by eliminating voluminous organelles like the nucleus and mitochondria.

However, throwing out the cell's nucleus comes with a price. When an immature red blood cell discards its nucleus, it throws the entire DNA coding instructions out the window too! Without DNA, there are no instructions to make replacement parts. As the cell infrastructure gets worn out, nothing can be replaced. Consequently, red blood cells have a limited lifespan – about four months – before they lyse (burst) from brittleness. Think of it like buying a brand new car, but never being able to replace the oil, tires, or brakes. That new car won't last very long without maintenance.

Without mitochondria, there will be no efficient and rapid production of ATP through aerobic respiration. In this case, however, it is perhaps for the best. Red blood cells should not consume the O_2 they are meant to deliver to other parts of the body.

Place your smartphone or tablet over the following QR code to see three other examples of the *structure-function* relationship in cells.

Problem 6

A female's ovary is tasked with making and secreting lots of estrogen, which is a type of lipid. Therefore, an ovary cell has more _____ than a typical human cell.

- a) rough endoplasmic reticulum
- b) nuclei
- c) ribosomes
- d) cytoskeleton
- e) smooth endoplasmic reticulum

Problem 7

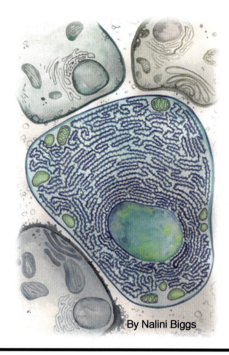

By Nalini Biggs

This painting is a color-enhanced reproduction of an image taken with an electron microscope. The blue cell is an activated **plasma cell** - a short-lived immune cell that synthesizes and secretes thousands of antibodies (which are large proteins) per second. What organelle is particularly prominent in the plasma cell? How does this prominence tie into the *structure-function* theme?

Problem 8 *Challenge!*

This cell lines the wall of the small intestine. The job of this cell is to grab as many nutrients as possible from the digested food that is flowing by, transport it into the cell and then out the other side (direction of nutrient flow shown by yellow arrows).

Examine this image.

What specializations do you see that optimizes this cell's function?

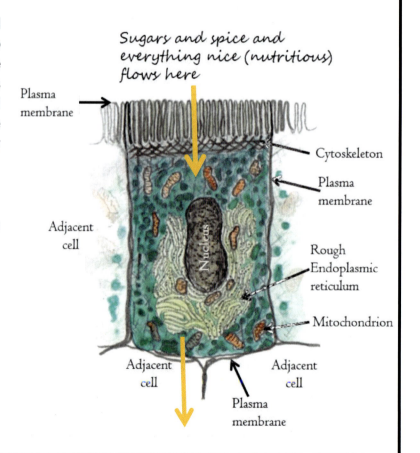

A collection of cells and their local extracellular products (i.e. anything they secrete that stays near the cells) makes a **tissue**. The study of cells and tissues, which is an important component of *Human Anatomy*, is most often achieved with a microscope. This branch of anatomy is histology.

Histology is an acquired taste. Many anatomy instructors *LOVE* histology (I'm no exception), but that love is shared by few students. There are several reasons for this discrepancy, and students' struggles to understand what they are viewing on a microscope slide is certainly one of them. The following example will illustrate.

As part of my research for this book, I surveyed 160 students during their fifth week of a *Human Anatomy* course. Students were shown a specific histology image and then asked to agree or disagree with this statement, "*I am comfortable studying this [histology] image with no textbook for guidance.*" Of the 160 students that responded, only 32% (n = 51) agreed or strongly agreed with the statement. Moreover, 10 of those 51 students failed at a basic and absolutely necessary skill in histology – to discern between cells and extracellular matrix. This is like saying "*I'm comfortable with driving*" but then pointing to the emergency brake when someone asks you were the gas pedal is.

It is my hope that the rest of this chapter will guide you to a pleasant, meaningful and practical approach to histology. For histology is wonderful once you learn how to study it, with new understandings revealed almost every time you peer into a microscope.

When first tasked with a histology slide, the usual first step is to look at a picture of the tissue in a textbook, a lab book or on the web before viewing the actual slide. But what you see in a book (or on the web) probably won't match your actual slide exactly. **Figure 6A** is a classic drawing of *loose connective tissue* that is usually presented in textbooks; **Figure 6B** is a photo of a slide entitled "*Areolar Loose Connective Tissue*" (at 100x magnification). While there are some similarities, the two images are clearly not identical.

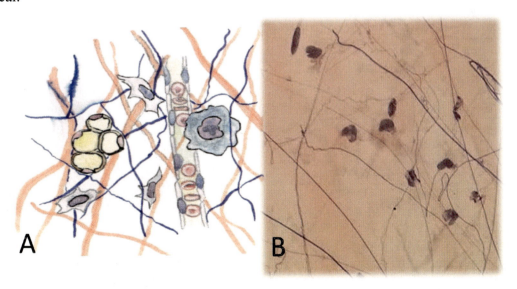

Figure 6. **A.** Schematic drawing of loose connective tissue showing the various elements that can be found in loose connective tissue. A sketch of this type and detail is standard fare in *Human Anatomy* textbooks. **B.** Photo taken of areolar connective tissue (magnified 100x), a type of loose connective tissue.

Why is this noteworthy? Because students often can't move past the disconnect between what they see *in a picture* and what is actually on *a prepared slide*. So, keep mulling that idea over in the front of your mind[4], while I move onto some basic histology guidelines.

As with every topic you encounter in anatomy, you should always try to identify the structure-function relationship.

 Before viewing a slide, try to answer this question first: *what's the function of this tissue?*

For without any concept of function, the tissue structure <u>won't make sense</u>. You risk attempting to memorize without understanding, defeating the purpose of the assignment.

So, start with a defined function. If you don't know, read about it in a text, the web, or ask a peer or instructor. What's important is that you have a function conceptualized BEFORE looking at that slide under a microscope. Problem 9 gives you some practice at identifying basic functions.

Problem 9

Match the following tissue samples with their function. You may have to use a textbook or other resource if you are unfamiliar with the tissue. As an example, one has been done for you.

Tissue	Function
adipose	create a boundary between two areas
skeletal muscle	store large quantities of fat
blood	transports stuff around the body
cartilage	generate movement of bones
epithelial	provides structural support but more flexible than bone (matched to cartilage)
small intestine	absorbs small nutrients from the food

Once you have an idea of the tissue's function, you can begin thinking about what types of structures could make that happen. I'll start with an easy example.

Adipose tissue has one major function - to store large quantities of fat. **Adipocytes** are cells that are dedicated to storing fat. To maximize fat-storing capacity, it would make sense to build adipose tissue almost entirely out of adipocytes (**Figure 7**).

---

[4] The place of mental processing, such as reasoning, is done in the frontal lobes of your brain, near your forehead. The phrase, "keep that in the back of your mind" as you contemplate some decision is anatomically incorrect.

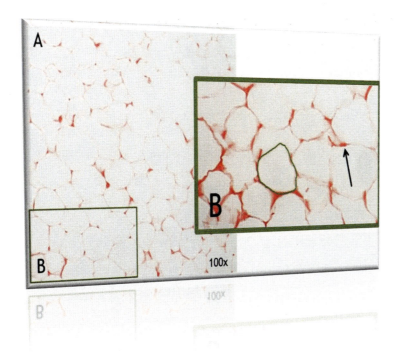

Figure 7.

A. Adipose tissue on a slide demonstrating the dominance of adipocytes.

B. Green inset. The plasma membrane of a single adipocyte is outlined in green. Each adipocyte is filled with fat (transparent) and therefore the nucleus is pushed to the outside (arrow indicating dark red oval).

In addition to the adipose tissue structure reflecting the function, adipocytes themselves are optimally structured for their function. The cytoplasm of an adipocyte is almost entirely devoted to fat storage. The fat is translucent, and therefore the cells have the appearance of being large, "empty" bubbles.

The following tutorial will introduce a few other themes and techniques associated with histology. After watching the tutorial, try your hand at some problems.

Problem 10

In this slide of loose connective tissue, identify a <u>cell</u> and an <u>extracellular fiber</u>. Answer questions 1 through 4.

1. What is the *function* of the cell you identified?

2. Name the cell you identified.

3. How many cells are in this image?

4. Name the fiber you identified.

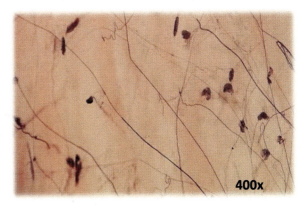

Problem 11

Two different magnifications of the epithelial tissue lining the esophagus (the tube that connects the throat to the stomach) are shown. The epithelial tissue is underlined in green.

1. What is the general function of epithelial tissue?

(Circle the correct answer to complete the sentence)

2. Based on the structure of this epithelial tissue (shown @ 400x magnification), its function is to

   a) permit rapid transport of substances between the environment (white space) and underlying cells.

   b) protect the tissue underneath the epithelial tissue from trauma.

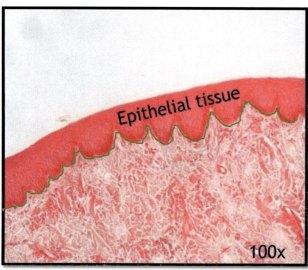
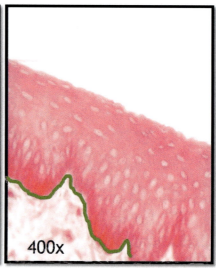

Problem 12

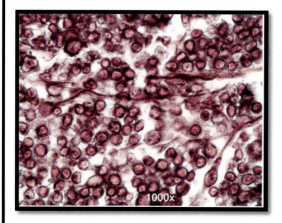

Lymph nodes are organs that hold large quantities of immune cells with reticular fibers. The word *reticular* means an intricate network. This image is a 1000x view of a lymph node.

1. Identify an immune cell and a reticular fiber.

2. Apply your reasoning skills to suggest a possible function for reticular fibers.

Dense connective tissue is densely packed with collagen fibers. The collagen fibers can be oriented in parallel (dense *regular* connective tissue). Alternatively, the collagen fibers can be grouped into smaller bundles with each bundle oriented a different direction (dense *irregular* connective tissue). The differences in *dense regular* and *irregular* tissue reflect differences in function. One type of dense connective tissue provides excellent support against being pulled from one direction, like a **tendon** (connects a muscle to a bone). The other type of connective tissue resists tearing when pulled in many possible directions, like skin. Use this information to solve Problem 13.

Problem 13

Each question has two correct answers.

1. Slide A is
   a) dense regular connective tissue.
   b) dense irregular connective tissue.
   c) from a tendon.
   d) from skin.

2. Slide B is
   a) dense regular connective tissue.
   b) dense irregular connective tissue.
   c) from a tendon.
   d) from skin.

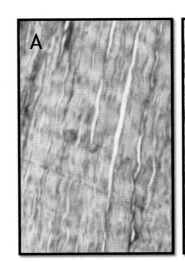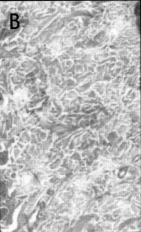

Regional anatomy slides – slides that are from a specific area of the body – often include multiple tissue types. To successfully navigate the slide and "find your bearings" (i.e. figure out what the heck you are looking at), you must be able to separate out the different tissue types. The successful identification of one layer or region can act as a landmark (or more appropriately, "body"mark), which is often useful in figuring out what is in the vicinity of that layer.

The histology of the small intestine serves as an example. The small intestine is a long tube that is designed to absorb small nutrients – sugars, amino acids, fatty acids, salts and water. We will use the small intestine as an example of delineating tissue layers in an anatomical region or organ, and to reinforce the *structure-function* theme.

The intestinal tract is a long tube. Digested food travels in the **lumen**, or hollow space in the center of the tube. If you cut the intestinal tract in a cross-section and look at the intestinal wall under a microscope, four layers are evident. The layers are (from internal lumen to external): *mucosa, submucosa, muscularis externa,* and *serosa*. The **mucosa** is the innermost layer; the name reflects its ability to secrete a lubricating mucous, which helps keep any food matter from abrading the intestinal epithelium. In the small intestine, the mucosa is highly folded to increase the surface. The internal boundary of the mucosa is a thin layer of muscle called the **muscularis mucosa.** *Muscularis*, of course, refers to muscle. Hence, this small layer of muscle (part of the *mucosa*) helps keep the shape of the many mucosa folds. The

submucosa is the next layer underneath. Indeed, the prefix *"sub-"* means below. Thus, the *submucosa* is the layer underneath the *mucosa*. The submucosa often carries blood vessels and may be rich in circular glands. Underneath the submucosa is a very thick layer of smooth muscle, called **muscularis externa**. As the name implies, this layer of muscle is more external (relative to the muscularis mucosa). The muscularis externa is actually further divided into two layers; the two layers have muscle fibers oriented perpendicular to each other. Their contractions collectively keep food moving through the small intestine. The outermost layer is the thin **serosa**, which secretes a slippery "serum" into the abdominal cavity to prevent friction as the coiled intestine rubs against itself and nearby organs.

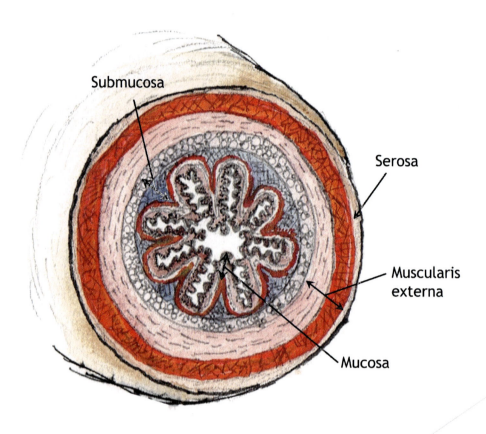

Figure 8. A cross section of the small intestine with the four layers color coded and labeled. A similar drawing or sketch is standard in an anatomy textbook.

Since the function of the small intestine is to absorb as many nutrients as possible as the intestinal juice (i.e. liquefied and digested food) flows by, then it makes sense that the epithelial tissue would be as expansive as possible. Increasing the surface area of the epithelial tissue increases the number of possible contact points between the epithelial cells and the nutrients, and therefore increases the number of nutrients that are absorbed. Not sure what I mean? Here's an analogy. Let's say 1000 fruit flies were being released into a classroom for 15 minutes and you were tasked with capturing as many flies as possible in those 15 minutes. You have unlimited access to sticky flypaper. How much flypaper would you layout? The answer should be *as much as possible*. For increasing the surface area of flypaper increases the likelihood that a fruit fly come in contact with flypaper. In the same way, increasing the surface area of the intestinal wall increases the rate of contact with the nutrients, thereby increasing the rate of absorbing those nutrients.

Problem 14

This is a cross section of a small intestine slide. Answer the following questions.

1. Identify layers A through E. You will need to reference both **Figure 8** and textual information presented in the previous two pages.

    A. _____
    B. _____
    C. _____
    D. _____
    E. _____

2. Use a pen or marker to trace ⌐⌐ the epithelial tissue on this image.

3. Relate the surface area (i.e. coverage) of the epithelial tissue to the function of this organ.

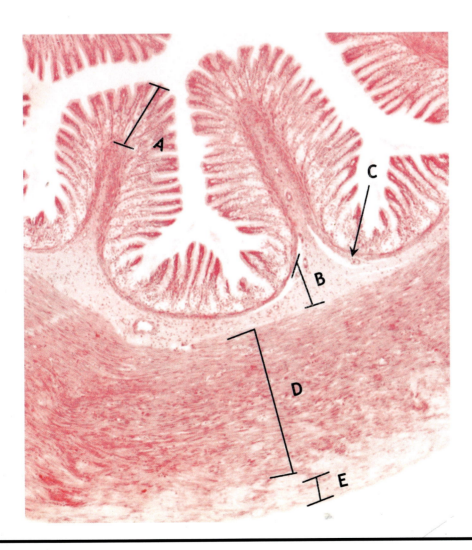

Chapter 8: Answer key

1. 1) nucleus   2) a   3) A eukaryotic nucleus is larger than a prokaryotic cell as indicated by the fact that structure A is larger than a single little dot (a single cell) in the cluster of region B   4) d.

2. 1) nucleus   2) nuclear pore   3) nuclear envelope   4) rough endoplasmic reticulum (ER)   5) smooth endoplasmic reticulum (ER)   6) peroxisome   7) Golgi body/apparatus   8) cytoskeleton   9) lysosome   10) mitochondrion (singular of mitochondria)   11) plasma (cell) membrane   12) ribosome   13) vesicle

3. plasma membrane, endoplasmic reticulum, Golgi body or apparatus, lysosomes, nuclear envelope

4. mitochondrial matrix, intermembrane space

5. *Each cell structure's function is listed on the same line.*

Smooth endoplasmic reticulum	synthesizes lipids
Nuclear envelope	membrane that surrounds chromosomes
Mitochondria	synthesizes ATP from energy stored in biomolecules
Ribosomes	reads mRNA instruction, builds (poly)peptide
Rough endoplasmic reticulum	folds polypeptides into proteins
Cytoskeleton	internal structure and highway system for vesicles
Vesicles	transports stuff among the endomembrane system
Golgi body	"post office"; ships out packages of molecules
Lysosome	breaks down, recycles particles and old molecules
Peroxisome	neutralize hydrogen peroxide, break down big fats
Nuclear pore	allows mRNA to exit nucleus

6. e

7. **Rough endoplasmic reticulum** is prominent in an activated plasma cell, encompassing almost all available cytoplasm. A plasma cell is an antibody-producing factory. Therefore, we expect the plasma cell to be enriched with ribosomes, which create polypeptides. Because (1) the proteins are large and thus require help folding, and (2) the proteins are ultimately going to be secreted and thus must be part of the endomembrane system, the ribosomes are attached to endoplasmic reticulum rather than being free ribosomes. Recall that endoplasmic reticulum helps fold and modify polypeptides into functional proteins; ER is also connected to the Golgi body and plasma membrane by way of vesicles. Therefore, the pathway for the antibodies to be secreted out of the cell is rough ER → Golgi body → plasma membrane.

8. The most obvious specialization is that the plasma membrane that faces the food is highly folded, creating a look of spiky hair. The highly folded plasma membrane increases surface area. As explained later in the chapter, increased surface area enhances the rate of nutrient absorption. In addition, we see a large number of mitochondria in the cell, which means the cell is energetically expensive. Because food particles are constantly being processed and transported throughout the cell, it makes sense that the transportation cost (i.e. energy spent moving vesicles and molecules) is high. Thus, extra mitochondria provide for the extra fuel.

9. *Each tissue's function is listed on the same line.*

adipose	store large quantities of fat
skeletal muscle	generate movement of bones
blood	transports stuff around the body
cartilage	provides structural support but more flexible than bone
epithelial	create a boundary between two areas
small intestine	absorbs small nutrients from the food

10. The only cell type in this picture is a fibroblast; its function is to synthesize and secrete (blast out) fibers. Based on the number of stained nuclei visible in this image, there are between 15 – 18 cells. The only clearly visible fiber type is the elastic fiber (dark, thin line), though you may be able to see a collagen fiber outlined.

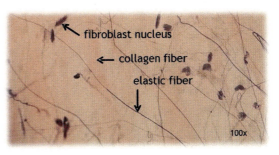

11. 1) Epithelial tissue provides at least one connected sheet of cells to form a boundary between two different areas within the body, or between the body and the external environment.

    2) b. *Explanation:* the structure of the epithelial tissue shows many layers of cells rather than a single layer. Multiple layers of cells indicate the primary function is to provide a defensive layer against possible trauma to the underlying tissue.

12. 1) Each black arrow points to a reticular fiber. Each blue circle encompasses one (of many) immune cells.

    2) Based on the provided description of a lymph node, the definition of the word reticular and this histology image, you should be able to deduce that the function of reticular fibers is to create a network that holds cells in place.

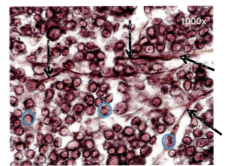

13. 1) a, c   2) b, d

14. 1) A. mucosa;   B. submucosa   C. muscularis mucosa D. muscularis externa   E. serosa

    2 + 3) The epithelial tissue is *partially* traced with a green line in this image. As you can see by the tracing, the surface area of the epithelial tissue is extensive. This structural specialization (lots of folding to increase surface area) maximizes the rate of contact between the epithelium and the digested biomolecules, ions and water. Increasing the rate of contact increases the amount of nutrients that are absorbed, which is the function of this organ.

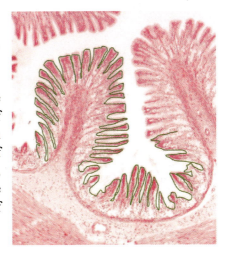

Chapter 9: Plasma membrane and lipids

They get along like oil and water.

The plasma membrane is the boundary between the cell's internal (**intracellular**) environment and the external (**extracellular**) environment. Any substances going in or out of the cell must pass through the plasma membrane. Therefore, the plasma membrane is responsible for regulating transport of substances between the cell and the extracellular environment. The plasma membrane is also tasked with receiving external stimuli or signals and initiating a response within the cell. For example, the plasma membranes of certain neurons are equipped to detect stimuli such as sound, touch or heat, and then transmit that information to the rest of the cell. Molecules involved in cell-to-cell communication (e.g. hormones) may also detected by the plasma membrane.

All cellular membranes are rich in lipids. Lipids are the fourth class of biomolecules. Unlike carbohydrates, nucleic acids and proteins, there is no common monomer for the lipids. Rather, **lipids** are a hodge-podge collection of biomolecules that are built from mostly hydrogen and carbon atoms. Consequently, the following rule of thumb applies.

 Lipids are generally nonpolar and hydrophobic biomolecules.

Three types of lipids are important to your studies: **fats & oils**, **steroids**, and **phospholipids**. Each type of lipid plays a different, but vital role in living organisms. Let's first discuss the three types of lipids and their uses in humans and/or infectious microbes before moving on to a more careful discussion of the plasma membrane.

Fats and oils are nonpolar molecules that are used predominately as an energy-rich source of fuel. The number of calories packed into a gram of fat is more than twice the number of calories in gram of sugar or protein. Thus, a fat molecule is an <u>efficient</u> way to store energy.

The term *oil* is used informally to describe a fat that is liquid at room temperature (**Figure 1**). Think about all the oils in your kitchen. You'll recognize them for their viscous yet liquid consistency. Compare oils to animal fat; the latter hardens into a solid but malleable form at room temperature.

Figure 1. A. Lard – a typical animal fat – at room temperature. B. Canola oil at room temperature.

Fats and oils are both composites of three fatty acids attached to a glycerol molecule (**Figure 2**). I like to think of glycerol as a clothes hanger, and each fatty acid as a long hydrocarbon rope hanging off of the glycerol "hanger".

Each **fatty acid** is itself a long hydrocarbon chain with a carboxyl group at one end. In a living system, the carboxyl group acts as an acid; this is why the entire molecule is referred to as a fatty *acid*. Fatty acids vary in the number of carbon atoms in their skeleton. For example, a fatty acid may have 12 or 15 carbon atoms. Some fatty acids are **saturated** – there are only single covalent bonds in their chain. **Unsaturated** fatty acids have at least one double covalent bond somewhere in their chain.

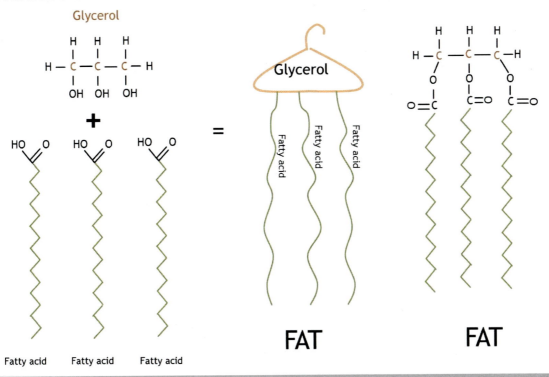

Figure 2. A fat or oil molecule has three fatty acids "hanging" off a glycerol molecule. The fatty acids have long hydrocarbon chains.

Problem 1

For each fatty acid, identify if it is *saturated* or *unsaturated*.

Then, determine the **number** of carbon atoms in each fatty acid.

	Saturated or unsaturated?	# of carbon atoms?
(top fatty acid)		
(bottom fatty acid)		

Problem 2

Draw out the full chemical structure of each fatty acid in the white space. Include all carbon and hydrogen atoms. Then add up the total number of hydrogen atoms in each molecule.

Total number of hydrogen atoms in the saturated fatty acid (top): _____

Total number of hydrogen atoms in the **un**saturated fatty acid (bottom): _____

Problem 2 demonstrates that a carbon skeleton with all single covalent bonds (e.g. a saturated fatty acid) has more hydrogen atoms than a carbon skeleton with one or more double covalent bonds.

Indeed, the qualifier *saturated*, when referring to the fatty acid, means that the carbon skeleton is "saturated" with hydrogen atoms. More information about the terms *saturated* and *unsaturated* and how this affects fat structure can be found by placing your smartphone or tablet over the following QR code.

Problem 3

The molecule on the *left* is (select BOTH correct answers)

a) a saturated fat.
b) an unsaturated fat.
c) liquid at room temperature.
d) solid at room temperature.

The molecule on the *right* is (select BOTH correct answers)

a) a saturated fat.
b) an unsaturated fat.
c) liquid at room temperature.
d) solid at room temperature.

We now turn to a second type of lipid – steroids. *Steroids* is a "hot button" word and invokes images of athletes with unnaturally large muscles. But the real definition of steroids is more mundane. A **steroid** is a biological molecule defined by a carbon skeleton arranged into the four-fused ring structure shown in **Figure 3**.

The most common steroid in your body is cholesterol. Yep, cholesterol. Hardly the molecule people think of when envisioning juiced athletes who cheat.

On a side note......

When eating more calories than you burn, your body converts the excess food into fat molecules, which is then stored in adipose tissue. Yet, fats are more than a "piggy bank" of extra energy. Fats also (1) provide insulation to conserve body heat, and (2) cushion joints and sockets. For example, the back of your eyeball is loaded in fat as space filler and provides soft cushioning for your eye in the bony socket. Without that fat cushion, our eyeball would risk constant friction, abrasion and jiggling in our skull.

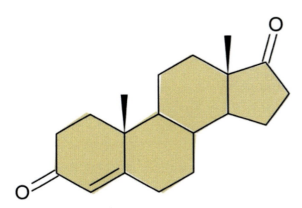

Cholesterol has two major roles in your body.

1. Cholesterol is part of the human plasma membrane.

2. Cholesterol in your body is chemically modified into hormones.

Figure 3. A steroid molecule demonstrating the four-fused ring structure that defines steroids.

Cholesterol's participation in the plasma membrane is saved for later in this chapter. Let's turn our attention to hormones built from steroids

Problem 4

This is cholesterol. Circle the only part of cholesterol that is polar. Color in the four-fused ring structure.

109

Problem 5

This structure is the shorthand version of a steroid hormone. Use the techniques from Chapter 4 to draw out the complete chemical structure of this steroid hormone in the white space. Include all of the carbon and hydrogen atoms.

Hormones that are built from cholesterol include some familiar names, such as estrogen, progesterone, cortisol and testosterone. These are **steroid hormones**. Each steroid hormone retains the four-fused ring structure but varies slightly in the side chains, which are organic structures protruding from the rings. Some of these steroid hormones are classified as **anabolic**. Anabolic steroids induce the body to synthesize, or build, muscle mass (**Figure 4**). The classic anabolic steroid is testosterone. When athletes talk about "steroids", they are referring to anabolic steroids, which is testosterone or some derivative of testosterone. For decades, professional athletes seeking a competitive edge have taken supplemental testosterone. Today, testosterone supplements (or any equivalent anabolic steroid) through injection, patches, or creams are considered cheating in many sports. Consequently, the use of extra testosterone or any steroid with potential anabolic effects are banned in most professional sports.

When biologists talk about steroids, they usually mean any biological molecule that is chemically classified as a steroid. A biologist using the word "steroid" could be referring to cholesterol or estrogen just as readily as testosterone. Clearly, it's important to have context when discussing the topic of *steroids*.

> *On a side note......*
>
> In 1998, professional baseball player Mark McGwire hit 70 home runs in the season, breaking the previous record of 61 home runs set in 1961. In 2001, Barry Bonds hit 73 home runs, breaking Mark McGwire's record.
>
> For most sports, athletic talent improves each decade. Yet in 2014, the most home runs hit by any player was only 40. Why the huge drop off in home run production? Because Major League Baseball (MLB) came down hard on illegal steroid use.
>
> In 2003, MLB implemented a testing program for banned substances, with anabolic steroids as their #1 target. High profile players began to test positive for anabolic steroids and were slapped with heavy fines, suspensions and public shame. Thus, fewer players are taking testosterone and consequently, the average bicep size (and number of home runs) in MLB slimmed down to more normal levels by the late 2000s.

Figure 4. Both males and females make testosterone. However, adult males produce about 10-20x more than adult females. This leads to sexual dimorphism (differences between males and females) in muscle mass, as demonstrated by the bicep girth of Mr. and Mrs. Gonzalez, both of whom do bicep curls regularly.

Phospholipids are the third type of lipid. Despite the fact that phospholipids get less publicity than either steroids or fats, phospholipids are incredibly important lipids; phospholipids are the basis of all membranes.

Phospholipids are distinguished by their "dual personality" in an aqueous environment. The *very* **hydrophilic head** includes a negatively charged phosphate group (hence the prefix *phospho-*). Branching off the head are two fatty acid **tails**; the tails are quite **hydrophobic** (**Figure 5**). So, we have the classic Dr. Jenkyl and Mr. Hyde complex in one molecule. Expose a phospholipid to water; how will it react since the head is *loving* water and the two tails hate it? Place your smart phone or tablet over the QR code on the left to find out.

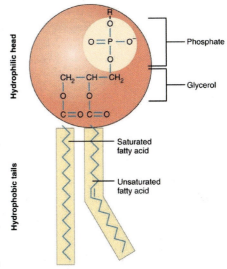

Figure 5. A phospholipid.

Image by OpenStax College [CC BY 3.0 (http://creativecommons.org/licenses/by/3.0)], via Wikimedia Commons.

Problem 6

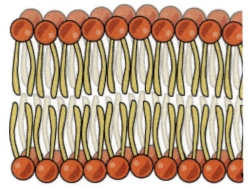

In this phospholipid bilayer, circle the region(s) that are hydrophilic and put a box around the region(s) that are hydrophobic.

Image by OpenStax College [CC BY 3.0 (http://creativecommons.org/licenses/by/3.0)], via Wikimedia Commons.

Problem 7

Which type of chemical interaction holds a phospholipid bilayer together?

    a) covalent bonds between the phospholipid heads

    b) ionic bonds between the tails of one phospholipid and the head of another phospholipid

    c) hydrogen bonds between the tails of the phospholipids

    d) hydrophobic interactions between the phospholipid tails

    e) All of these chemical interactions are involved in holding the phospholipid bilayer together.

Problem 8 *Challenge!*

The presence of water is important to keeping a phospholipid bilayer intact. Describe what would happen, if anything, if an existing phospholipid bilayer were suddenly surrounded by oil instead of water.

A phospholipid bilayer is at the "heart" of every membrane. And to be functional, the membrane must maintain fluidity. *What exactly does it mean to be fluid?* It means the membrane molecules, especially the phospholipids, must be able to mill about their leaflet (layer) easily.

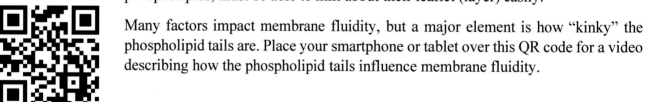

Many factors impact membrane fluidity, but a major element is how "kinky" the phospholipid tails are. Place your smartphone or tablet over this QR code for a video describing how the phospholipid tails influence membrane fluidity.

Membrane fluidity is <u>important</u> because it

    (1) allows other molecules embedded in the membrane to move, and

    (2) temporarily creates small gaps that allow certain small molecules to cross the bilayer.

Problem 9

Phospholipids in human plasma membranes maintain fluidity by _____ which affords the phospholipids just the right amount of "dancing" room.

a) having two saturated tails
b) having one unsaturated and one saturated tail
c) having all unsaturated tails

Phospholipids are not the only lipid in human plasma membranes. Cholesterol molecules are embedded as stabilizers. Without cholesterol, the phospholipids in the bilayer actually move too much. The result would be a hyper-fluid membrane with increased permeability, allowing too many molecules to slip through the temporary gaps created during phospholipid movement (**Figure 6**). Cholesterol is inserted to reduce some of the permeability, which is necessary for the plasma membrane to have some selectivity about which molecules (or ions) move in and out of the cell. The plasma membrane's **selective permeability** is discussed at length in Chapter 11.

Phospholipid bilayer

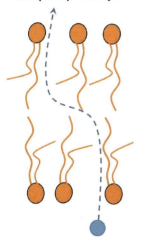

Lipid bilayer with cholesterol

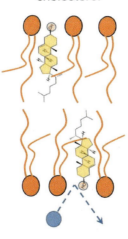

Figure 6. A phospholipid bilayer without cholesterol is too fluid for a human cell membrane and would allow too many molecules (blue circle) to travel through the many transient gaps. The addition of cholesterol is the "stop-gap" measure (as in actually stopping up the gaps created by the kinky, mobile phospholipids). Notice the orientation of cholesterol in the membrane; the only polar region (a hydroxyl group) aligns with the hydrophilic phospholipid heads. Cholesterol's four-fused rings align with the hydrophobic tails.

Problem 10

Which membrane has the lowest permeability (i.e. allows the fewest variety of molecules to cross the lipid bilayer)?

A.

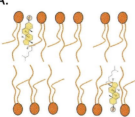

B.

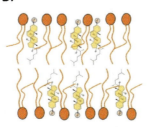

C.

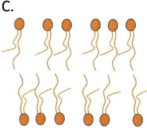

All membranes start with a lipid bilayer, but each membrane is specialized by its protein and carbohydrate enhancements. A variety of embedded proteins are prominent in the plasma membrane, constituting about half of the plasma membrane by weight. Chains of sugars adorn the outer leaflet (layer), attached to either lipids or proteins (**Figure 7**).

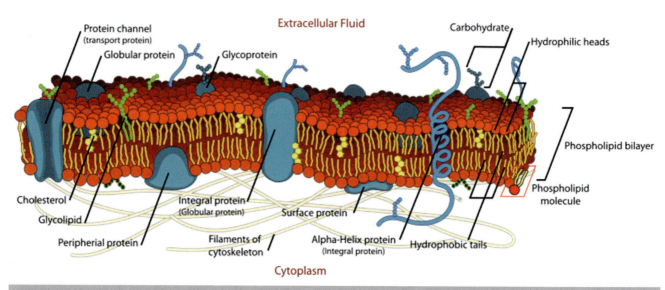

Figure 7. Plasma membrane. The lipid bilayer includes cholesterol, a variety of proteins (blue), and sugar chains (green) lining the top layer.

Membrane proteins – which are diverse– can be categorized by *structure* or *function*.

Structurally, a membrane protein is classified by how far it delves into the bilayer. Proteins that are thoroughly embedded into the bilayer are **integral proteins**, because they are *integrated* into the bilayer. In contrast, **peripheral proteins** lie on the *periphery* (or edge) of the bilayer (**Figure 8**). A very common subtype of integral protein is the **transmembrane protein**, which means the membrane protein crosses the entire bilayer and has regions peeking out of both sides of the bilayer.

What do these membrane proteins do? All kinds of really important things! Many membrane proteins are involved in molecular transport, transferring small molecules in and/or out of the cell. Some membrane proteins act as enzymes, catalyzing chemical reactions. Other membrane proteins are receptors for chemical signals that travel through the extracellular fluid. Membrane proteins can be structural, such as those that anchor cytoskeletal elements or extracellular fibers in place. A few membrane proteins *double dip* in function. For example, ATP Synthase (a transmembrane protein found in your inner mitochondrial membrane) is both membrane transporter and enzyme.

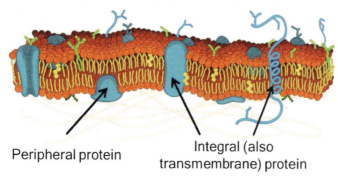

Figure 8. Membrane proteins are classified as *peripheral* or *integral*.

Problem 11

A membrane protein involved in molecular transport will also be a(n) _____ protein.

a) enzymatic   b) peripheral   c) structural   d) integral

Problem 12 *Challenge!*

What class of membrane proteins is most likely to exist as peripheral proteins?

a) membrane transporters   b) enzymes   c) receptors

On human cells, short sugar chains arranged in specific motifs protrude into the extracellular fluid. These chains are attached either to the fatty acid tails (**glycolipid**) or a membrane protein (**glycoprotein**). The prefix *glyco-* should remind you of *glyco*gen, which is our storage polymer of glucose. Here, *glyco-* still refers to a polymer of glucose, albeit a short, branched chain. The second part of the word describes what the sugar chain is anchored to. If the sugar chain is attached to fatty acids, it is known as a glyco<u>lipid</u>. If the sugars attach to a protein, then the entire structure is referred to as a glyco<u>protein</u>. Simple enough.

Glycolipids and glycoproteins serve as *barcodes* that create an I.D. or identify for our cells. They are found on the outer leaflet (layer) of the plasma membrane only for external visibility. Interestingly, it matters little whether these sugar mosaics are attached to proteins or lipids. The specific branching pattern of the sugars is what is important (**Figure 9**). For that reason, I collectively refer to them as glyco"tags".

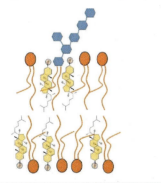 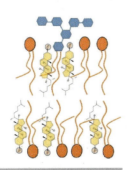

Figure 9. Two different glycolipids (blue) demonstrating how glucose chains can be arranged into different shapes or motifs.

Blood type serves as an example of the function of glycotags. Individuals with A type blood carry a specific glycotag motif on the surface of their red blood cells. Type B individuals display a different shaped glycotag. Type O individuals don't display either patterns exhibited by type A or B individuals. These red blood cell glycotags are barcodes for your immune system, which identifies the cells as *self* or *non-self* (i.e. foreign). For example, a person with A type blood recognizes the "A" glycotags as self, but the B glycotag is an unrecognized pattern that is deemed foreign.

This has profound implications in blood transfusions. As an example, I am O type blood. My immune system does not register either A or B glycotags as *self*. If I were to be infused with A (or B) type blood, my immune system would shift into overdrive and reject the transfused red blood cells. Such an infusion could be a fatal mistake.

On a side note......

Transplant rejections occur when a host (the recipient) receives an organ or tissue from a donor, but the host's immune system destroys the transplanted organ or tissue. This can happen when the donor cells display different shaped glycolipids or glycoproteins than the host. The host's immune system does not register these new glycotags as *self*, and therefore rejects the tissue or organ. The consequences of a rejection can be extreme, even fatal. For this reason, it is critical that the glycotags displayed on donor tissue match the recipient's glycotags as closely as possible.

We've now reviewed the structure of the plasma membrane and briefly explored the function of each component. Chapter 11 expands on the topic of membrane structure and function as it relates to molecular transport. For now, I want to leave you with one last thought.

The *structure-function* relationship applies to plasma membranes. Just like cells specialize, membranes specialize too. The membrane composition is not universal in a human body; each cell selects specific membrane proteins to insert, the amount of cholesterol to integrate, and even which glycotag variants to display.

Moreover, the plasma membrane can change across a single cell! Neurons are great examples of this. One end of a neuron might have membrane transporters specific to allowing Na^+ to pass; the other end of the same cell is rich with different protein transporters specific for Ca^{2+}. Remember to apply the *structure-function* concept to membrane specializations in your Human Physiology course. We conclude this chapter with one last problem.

Problem 13

In this image, find and label the following membrane components:

- ✓ A phospholipid
- ✓ Integral/transmembrane protein
- ✓ Glycolipid
- ✓ Cholesterol
- ✓ Glycoprotein
- ✓ Extracellular fluid
- ✓ Intracellular fluid
- ✓ Hydrophobic region of the membrane
- ✓ Peripheral membrane protein

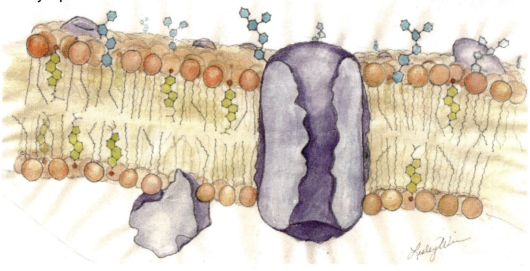

Chapter 9: Answer key

1. Top fatty acid – **saturated**; 12 carbon atoms
   Bottom fatty acid – **unsaturated**; 10 carbon atoms. *See Chapter 4 for more guidance on how to calculate carbon atoms.*

2. Saturated fatty acid – 22 hydrogen atoms; unsaturated fatty acid – 20 hydrogen atoms.

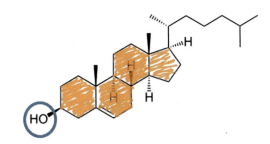

3. Left – b, c   Right – a, d

4.

5. The molecular formula is $C_{19}H_{26}O_2$. Watch the following video for a tutorial on how to draw the complete chemical structure.

6.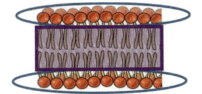

7. d

8. The phospholipid bilayer is held together by the hydrophobic tails coalescing out of fear of water. If oil was used in place of water, the tails would have nothing to fear and therefore would not strongly aggregate. The phospholipid bilayer would probably disintegrate as individual phospholipids cavort with the fats rather than with each other.

9. b

10. b
    Explanation: The more cholesterol in a membrane, the lower the permeability. Cholesterol molecules make it more difficult for the phospholipids to sway freely and create those transient gaps that make the membrane permeable.

11. d
    Explanation: If a membrane protein provides passage for a particular solute, then the protein needs to be fully integrated into the membrane to provide complete access between the intracellular and extracellular environments.

12. b
    Explanation: Enzymes are most likely to be peripheral proteins as they catalyze chemical reactions in a specific locale (such as the intracellular environment). Thus, they use the membrane as an anchoring spot while most of the protein is exposed on one side of the membrane. As described above, membrane transporters need to be integral proteins to function. Receptors are also usually transmembrane – one "face" sticks out in the extracellular fluid to receive a chemical signal; the other face pokes into the cell cytoplasm to relay the signal.

13.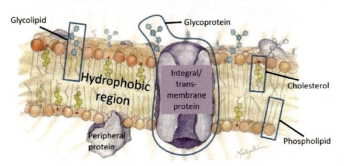

Chapter 10: Energy transformations

Better living through chemistry (and physics).

All living organisms require usable energy that is "spent" to do work. What work? This work:

- **movement** (e.g. muscle contractions, separating chromsomes during cellular division, powering sperm flagella),
- certain types of molecular **transport**, and
- powering some types of **chemical reactions**.

The objective of this chapter is to learn how energy is transferred from one form to another in order to drive cellular work. It's a complicated discussion, and one that invokes some physics. Physics is not my favorite subject, so the physics presented here is sugar coated.

First, we need to review some forms of energy that are applicable to living organisms. Energy is classified as either *kinetic* or *potential*.

Kinetic energy is energy of the "now". The *blink and you'll miss it* variety. Kinetic energy includes any sort of motion, light or heat. When kinetic energy is happening, it's happening now.

Potential energy is the energy in a "piggy bank". This form of energy is stored for later. Two significant types of potential energy are *chemical energy* and *concentration gradients*. **Chemical energy** is the energy stored in chemicals and their bonds. For our purposes, we usually refer to chemical energy as the energy in covalent bonds. Concentration gradients – where ions or molecules are concentrated into a particular space – are also a form of potential energy. When a cell taps into potential energy (i.e. spends it), the potential energy is transformed either into kinetic energy and/or another form of potential energy.

Problem 1

List three types of kinetic energy:

List two types of potential energy:

Which of the five forms of energy listed in Problem 1 are captured by organisms?

- ✓ **Plants** (and other photosynthetic organisms) capture **light energy**.
- ✓ **Animals** and **living pathogenic microbes** capture the **chemical energy** stored in carbohydrates, lipids, nucleic acids and proteins.

On a side note……

No organism captures heat as a source of energy to do work. For example, let's say you are running a marathon. At mile 13, your muscles fatigue because they are *running* (ha!) low on energy. How do you replenish your energy? You eat. You do not stand by a fire to "re-fuel" your muscles. If heat was a source of energy you could capture, then standing by a fire might have the same effect as eating 200 calories of food.

For humans and other warm-blooded mammals, maintaining a specific body temperature isn't the same as being able to capture heat from a system (i.e. a fire) and harness that heat to do work.

For example, the potential energy stored in gasoline is converted to the kinetic energy of heat and motion to move a car (moving a car is the "work"). Likewise, living organisms convert their fuel (e.g. *light* or *chemical energy*) to other types of energy to perform work. Humans and pathogenic bacteria extract the chemical energy from biomolecules by **cellular respiration**. The chemical energy released in cellular respiration is ultimately converted into motion, heat, concentrating a substance, and/or creating new covalent bonds in molecules. Plants and other photosynthetic organisms are built to capture a different type of energy – the kinetic energy of light. Light energy is then transformed into the chemical energy of sugars through a process known as **photosynthesis**.

We need to understand the rules that govern an organism's ability to transfer energy from one type to another. This is the domain of physics, so of course there are laws involved. The rules governing energy transformations are technically known as the *laws of thermodynamics*. The wording of each law does not lend itself to an easy interpretation. Therefore, a layman translation to how these laws apply to biological systems is helpful.

1st Law of Thermodynamics – Energy cannot be created or destroyed, only transformed.

What this really means: Money doesn't grow on trees; energy can't be produced out of thin air. You *can't* create energy. But if an organism can capture existing energy (e.g. light, chemical energy), then it can transfer that energy into some other form and usually perform work in the process. However, the total quantity of energy at the beginning of the transfer must equal the quantity of energy at the end of the transfer[1].

Let's say we run a chemical reaction. The chemical energy of the starting materials (**reactants**) is 10 calories. After the chemical reaction has occurred, the chemical energy stored in the covalent bonds of the ending chemicals (**products**) is 4 calories. However, the 1st law of thermodynamics states that energy cannot be destroyed, so the total energy at the end of the transformation needs to be 10 calories. Where did the other 6 calories go? The other 6 calories were released as heat – a form of kinetic energy (**Figure 1**).

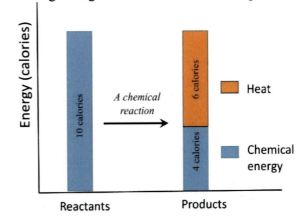

Figure 1. The total energy going into a transformation must equal the amount of energy coming out.

---

[1] There are many measurements of energy, including joules, watts, and calories. Calories is usually the most applicable to human physiology. A calorie (lowercase 'c') is the amount of energy required to raise 1 gram of water by 1°C. Nutritional labels list Calories with an uppercase 'C' which is = 1 kilocalorie or 1000 calories.

2nd Law of Thermodynamics – In an isolated system, entropy must stay the same or increase.

What this really means: In physics, entropy means a general state of disorder and it applies to forms of energy. For our purposes, the translation is this. Each energy transformation must increase the amount of disordered energy (i.e. energy that is unavailable for work). And in biological systems, the type of energy unavailable for work is the kinetic energy of **heat**. In summary, any time a cell makes an energy transformation, some of that energy is released as heat.

Key point! In any energy transformation, the starting energy must equal the ending energy (1st law) and some of that ending energy must be heat (2nd law). Therefore, it follows that <u>a chemical reaction can only proceed if the chemical energy of the reactants is higher than the chemical energy of the products</u>.

Problem 2

A chemical reaction can proceed *if and only if* it adheres to the two laws of thermodynamics. Consider the following four graphs charting the energy of four different chemical reactions. Evaluate each graph and determine if the chemical reaction satisfies BOTH laws of thermodynamics. As an example, the first one has been done for you.

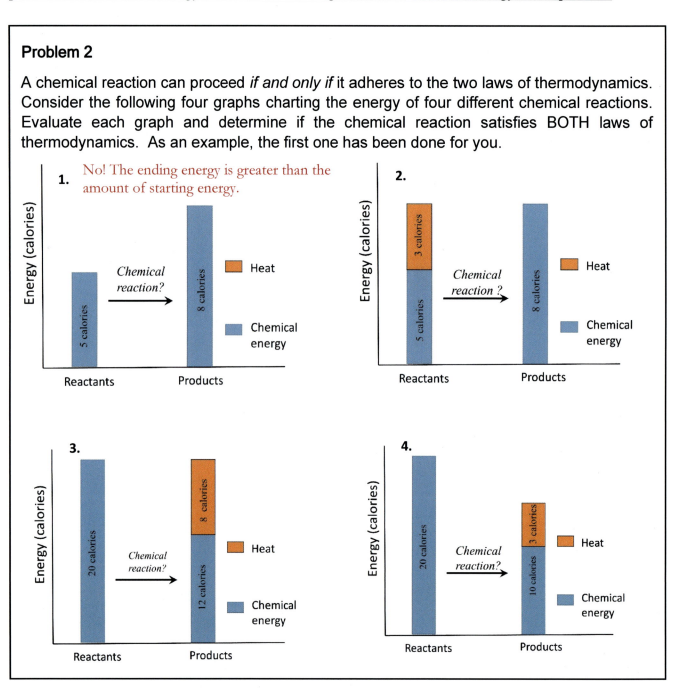

1. No! The ending energy is greater than the amount of starting energy.

Please review your answer to Problem 2 with the answer key at the end of this chapter before continuing.

Let's learn how cells tap into potential energy in order to do work. Remember there are two types of potential energy – chemical energy and concentration gradients. I'll focus first on chemical reactions, where chemical energy is involved.

Consider the following chemical reaction. *Can this chemical reaction even happen (based on the laws of thermodynamics)?*

$$\underbrace{2\,H_2 + O_2}_{Reactants} \xrightarrow{?} \underbrace{2\,H_2O}_{Products}$$

Based on the **key point** from the previous page, this reaction can happen *if and only if* the chemical "value" of the reactants is greater than the chemical "value" of the products. Well.....is the chemical energy stored in two hydrogen gas molecules and one oxygen molecule greater than the chemical energy of two water molecules? Let's discuss.

In my experience, most introductory biology textbooks emphasize these two key points.

(1) A chemical reaction <u>can</u> proceed if and only if the chemical energy of the reactants is greater than the products. Such a chemical reaction is called **spontaneous**.

(2) In every chemical reaction, some energy is lost as heat that cannot be recaptured (i.e. recycled) by an organism to perform additional work.

The same textbooks usually fail to divulge some critical information. Namely, whether or not the chemical energy of the reactants is greater than the products. That would be a key piece of information, yes? So let's get down to business and address that right now. Here's the rule of thumb you need.

 Nonpolar covalent bonds have *more* chemical energy than polar covalent bonds.

Let's invoke that rule of thumb to solve the above problem. *Do H_2 and O_2 have more chemical energy than H_2O?* H_2 is two hydrogen atoms joined by a single covalent bond. Since the covalent bond is shared by two "little sibling" hydrogen atoms, the covalent bond in H_2 is *nonpolar*. O_2 is also made of nonpolar covalent bonds.[2] In contrast, water molecules consist of polar covalent bonds. Based on the rule of thumb, we deduce that the reactants must have more potential energy than the products, and therefore the chemical reaction CAN happen (i.e. it's **spontaneous**[3]).

So you see, if the reactants have more nonpolar covalent bonds than the products, we expect the reaction to follow the two laws of thermodynamics and therefore be capable of happening. Such reactions are labeled *spontaneous* or *energetically favorable*; some heat is always released during the reaction.

---

[2] Review Chapter 2 if you are unsure as to why H-H and O_2 have nonpolar covalent bonds.
[3] The term *spontaneous* should not be confused with instantaneous. A spontaneous chemical reaction is one that <u>can</u> happen, not necessarily one that <u>will</u> happen right this millisecond.

Problem 3

Use the information in the previous two pages to determine if each chemical reaction is spontaneous or not.

1. $C_6H_{12}O_6 + 6\ O_2 \rightarrow 6\ CO_2 + 6\ H_2O$

2. $6\ CO_2 + 6\ H_2O \rightarrow C_6H_{12}O_6 + 6\ O_2$

3. 2 H-C-C-C-C-C-C-C-C-H (octane, all H) + 25 $O_2 \rightarrow$ 18 H_2O + 16 CO_2

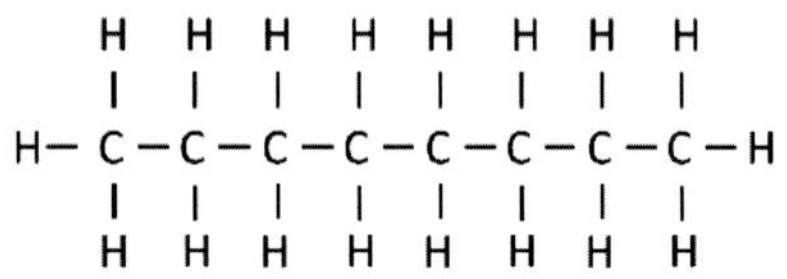

4. $N\equiv N + 3\ H_2 \rightarrow 2\ H-NH-H$

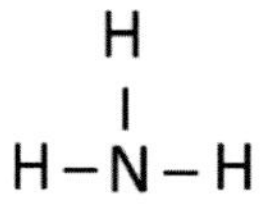

Problem 4

Here are the chemical structures of glucose, an amino acid, and a fatty acid – three organic molecules commonly used as fuels in our body. Which molecule has the most chemical energy *per gram*? How do you know?

Polar amino acid   Glucose 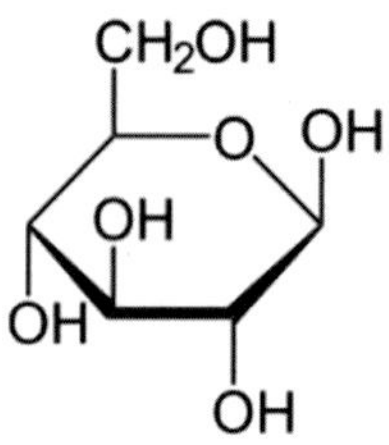   Fatty acid

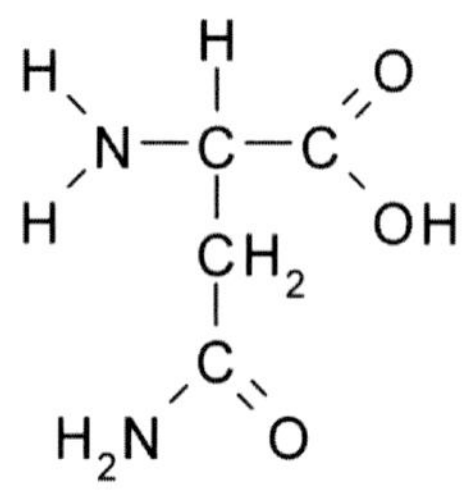

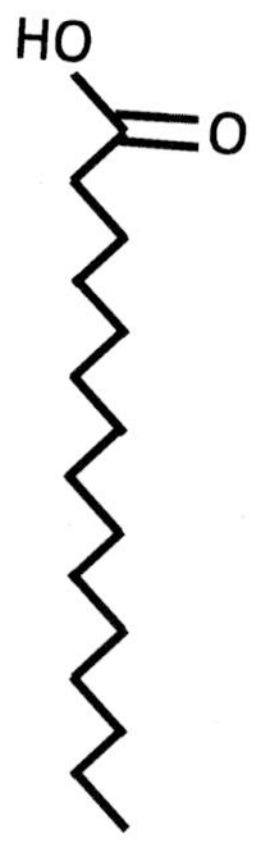

Please compare your answers to Problems 3 and 4 with the answer key at the end of this chapter before moving on.

Nutritionally, proteins (built from amino acids) and sugars each release about 4 kilocalories of energy per gram. Yet a gram of fat has more than twice as much energy (9 kilocalories/gram). Many students of nutrition memorize this fact without questioning why. But now you know why. Fats are almost all nonpolar covalent bonds, and because nonpolar covalent bonds have more chemical energy than polar covalent bonds, fats have the highest caloric value per gram.

Any chemical reaction that is spontaneous must release heat. Such reactions are termed **exergonic** (think of "*ex-*" as in *ex*pel the *ex*tra energy). The converse is an **endergonic** reaction, which is most definitely not energetically favorable. In an endergonic reaction, the reactants have *less* chemical energy than the products.

Despite what you read earlier, endergonic reactions do happen, but ONLY when additional energy is provided to the reactant side of the equation. *Huh?* Don't worry – I'll explain in a bit. First, let's get a better handle on the terms exergonic and endergonic.

Problem 5

Classify each reaction from Problem 3 as either exergonic or endergonic.

3-1: _____   3-2: _____

3-3: _____   3-4: _____

Use your smartphone or tablet to access the following video, which addresses the paradox about how an endergonic reaction can proceed while still adhering to the two laws of thermodynamics.

After watching the video, you know that an endergonic reaction can happen as long as the total energy input is greater than the chemical energy of the products. The video uses photosynthesis as an example, where the kinetic energy of light is added to increase the energy input to a level that satisfies both laws of thermodynamics. Yet, humans and pathogenic bacteria are not capable of transforming light energy into chemical energy. So……

How do humans (and pathogenic bacteria) run endergonic reactions?

The answer is by coupling (i.e. pairing) an endergonic reaction with a highly exergonic reaction. The energy released from a highly exergonic reaction provides the energy needed to drive the endergonic reaction. The most commonly used exergonic reaction involves ATP.

In Chapter 6, we learned the basic structure of ATP. ATP is a nucleotide with three phosphate groups (**Figure 2**). Access this video to learn just how ATP is used to drive many types of cellular work, including endergonic reactions.

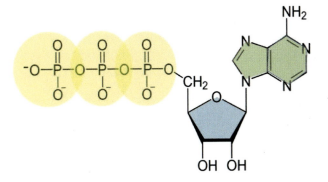

Figure 2. ATP. The **TP** stands for **triphosphate**, indicating the three (tri) phosphate groups attached.

The terms *endergonic* and *exergonic* are one pair of antonyms that describe chemical reactions. Another set of oft-used antonyms are *catabolic* and *anabolic*. A **catabolic** chemical reaction is one that takes a large molecule and breaks it down into smaller molecules. An **anabolic** chemical reaction is the opposite – taking smaller molecules and building something larger. One way to remember this is *cata*bolic has the prefix *cata-* which reminds me of a *cata*strophe. In a catastrophe, something large and orderly (e.g. building, airplane) is destroyed into smaller parts (e.g. rubble). In a catabolic reaction, a large, ordered molecule is broken into smaller parts.

As an example, ATP → ADP + P is a catabolic reaction; the larger molecule (ATP) is broken into two smaller molecules. Textbooks or your professor may refer to this as *ATP catabolism*. Photosynthesis is an example of an anabolic reaction, where smaller molecules (CO_2 and H_2O) are combined to make a larger molecule (glucose).

Problem 6

Determine if whether each chemical reaction is catabolic or anabolic. As an example, the first one has been done for you.

1. $6\ CO_2 + 6\ H_2O \rightarrow C_6H_{12}O_6 + 6\ O_2$ *anabolic*

2. $C_6H_{12}O_6 + 6\ O_2 \rightarrow 6\ CO_2 + 6\ H_2O$ _____

3. ATP → ADP + P   _____

4. ADP + P → ATP   _____

5. 5 amino acids → peptide   _____

6. sucrose → glucose + fructose   _____

You might notice a correlation – catabolic reactions tend to be exergonic while anabolic reactions usually require additional energy. This trend isn't a hard and fast rule (hence no rule of thumb), but it is usually applicable to biomolecules. When a cell builds a polymer from monomers (a process that is anabolic), it spends energy to make it happen. Likewise, the catabolic process of taking any polymer and breaking it back into its individual monomers usually releases energy.

A fair number of chemical reactions transfer electron(s) from one atom or molecule to another. These are classified as *oxidation-reduction* reactions. A shortened term – *redox* – is customary (**Figure 3**).

Figure 3. The energetically favorable transfer of an electron from a sodium atom to a chlorine atom is an example of a **redox** reaction.

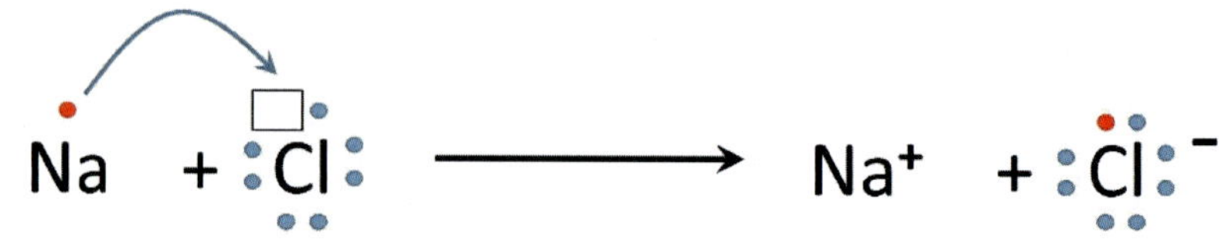

L E O	G E R
o l x	a l e
s e i	i e d
e c d	n c u
t i	t c
r z	r e
o e	o d
n d	n
s	s

Figure 3 shows a classic redox reaction where an electron is transferred from sodium to chlorine, making two ions in the process. You learned about this transfer in Chapter 2.

The atom or molecule that loses its electron is said to be **oxidized**. The atom/molecule that gains the electron is **reduced**. If you have a hard time keeping the definitions straight, one cutesy way to remember is with the saying LEO the lion goes GER[4]. In **Figure 3**, sodium lost an electron, so sodium is oxidized. Chlorine was reduced.

Problem 7

In this chemical reaction, what atom or molecule was oxidized? What atom or molecule was reduced?

$$2\ Ca^+ + Cl_2 \rightarrow 2\ Ca^{2+} + 2\ Cl^-$$

Like all chemical reactions, redox reactions must follow the laws of thermodynamics. In any redox reaction, there is a molecule or atom that is the electron donor (the oxidized agent) and a molecule or atom that is the recipient (the reduced agent). Their relative "desire" for that transferred electron determines if the reaction is energetically favorable or not. Here is the rule of thumb.

 A redox reaction is energetically favorable only when the recipient has a higher affinity (i.e. stronger desire) for the electron than the donor.

An example will illustrate this rule of thumb. In Problem 7, the chlorine atoms wanted another electron much more than calcium wanted its 19th electron. So, this redox reaction was energetically favorable. The electron moved from a donor (calcium) that didn't want the electron, to chlorine (the recipient) that desired the electron very strongly.

The term(s) oxidation or reduction can apply to reactions where there doesn't appear to be any transfer of electrons. The process of cellular respiration is a perfect example. In cellular respiration, glucose ($C_6H_{12}O_6$) is oxidized into CO_2 to make ATP energy. Yet the overall process (see Problem 3) does not alter the net charge of any molecule. *Confused as to how this could be a redox reaction?* You certainly would not be alone in your confusion. Ahhh….but the devil is in the details. And I wager that if you've made it this far into the book, you're smart enough to understand those devilish details. Watch the following tutorial to be enlightened about how cellular respiration qualifies as a redox reaction.

---

[4] Students often mistake the word "oxidized" as something requiring an oxygen atom. In fact, in chemistry, you might even learn that oxygen (O_2) is a great oxidizing agent – which it is. But don't confuse the two terms. If an instructor asks you which chemical has been oxidized, don't just automatically pick something with an oxygen atom or chances are, you'll get the answer wrong.

Problem 8

In this reversible reaction, two hydrogen atoms are added to turn pyruvic acid into lactic acid. Conversely, lactic acid becomes pyruvic acid when the two hydrogen atoms are removed. Which form (pyruvic acid or lactic acid) is the oxidized form? Which form is reduced? *Hint:* look at it from the middle carbon's perspective (yellow box).

Pyruvic acid ⟷ Lactic acid

Problem 9 *Challenge!*

In which direction is the chemical reaction energetically favorable?

  a) Pyruvic acid → Lactic acid   b) Lactic acid → Pyruvic acid

Let's return to ATP. Earlier in the chapter, we learned that ATP catabolism releases energy that can "pay for" an endergonic reaction and other types of work. If we think of ATP like cell money, then catabolism of ATP into ADP and P is analogous to *spending* that money. Obviously, a cell must continually renew its ATP supply (i.e. make more money) or it will die. Let's explore the processes that cells use to rejoin ADP and P into ATP. In the famous words of Cuba Gooding Jr., "SHOW ME THE MONEY!" (Jerry Mcguire movie reference).

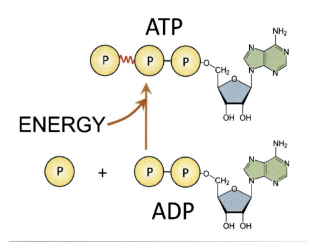

Figure 4. Reforming ATP through attaching a third phosphate group requires substantial energy "payment".

Making ATP is an uphill battle. Attaching that third phosphate group (PO_4^{3-}) is challenging because it is strongly repelled by the equally negatively-charged ADP. So forming that high energy covalent bond of ATP is going to take some serious energy compensation (**Figure 4**). Remember our two laws of thermodynamics. To run an endergonic reaction like ADP + P → ATP, additional energy input is required. What type of energy? That depends on what organism you are.

Humans and all living pathogens transform the *chemical energy stored in organic molecules* into the *chemical energy of ATP* through a series of chemical reactions and processes collectively known as **cellular** respiration.[5] Of course, not all of the chemical energy stored in these organic molecules is directly converted into the chemical energy of ATP; some of the energy is transformed into heat. However, that heat is still useful to humans as it helps keep our core body temperature around 98°F.

Plants transform the kinetic energy of light into the chemical energy of ATP as part of **photosynthesis**. A handful of microbial species tap into the chemical energy of <u>in</u>organic molecules that exist in a reduced state. Such microbes use the oxidation of these inorganic molecules to make ATP in a process known as **chemosynthesis**.

Problem 10

A bacterial strain performs this chemical reaction:

$$4\ Fe^{2+} + O_2 + 4\ H^+ \rightarrow 4\ Fe^{3+} + 2\ H_2O$$

This reaction is energetically favorable and performed by bacteria in oxygenated waters to create a reddish precipitate (right image).

1. The iron (Fe) atom was _____.
   a) catabolized
   b) anabolized
   c) oxidized
   d) reduced

Image of iron precipitates from iron bacteria courtesy of *www.nhep.unh.edu*

2. The bacteria couple the energy released from the above chemical reaction into the chemical energy of ATP. Therefore, this reaction is part of _____.
   a) photosynthesis
   b) aerobic respiration
   c) anaerobic respiration
   d) chemosynthesis

Metabolism describes chemical reactions in a cell. When a cell performs cellular respiration, photosynthesis, or chemosynthesis, the cell is actually running a sequence of chemical reactions that function similarly to an assembly line. For example, cellular respiration is not achieved in a single chemical reaction; it is a process that includes over twenty chemical reactions performed in a defined sequence. A step-wise series of chemical reactions – such as the *process* of cellular respiration – is known as a **metabolic pathway**.

At this point, we've covered energy transformations as it relates to chemical energy. I will now briefly detour to concentration gradients, the other type of potential energy.

---

[5] Aerobic respiration is a *type* of cellular respiration and indicates that O_2 is used to oxidize the organic molecule as part of the process. Some bacteria perform <u>an</u>aerobic cellular respiration, where some other molecule besides O_2 is used to drive the oxidation of the organic molecule. For humans, all cellular respiration is also aerobic respiration.

When the cell separates two compartments by a membrane and concentrates a substance in one of those compartments, you have a **concentration gradient**. Simply, one compartment has a higher density of some solute than the other. This is a form of potential energy that can be tapped to do work. Solutes that are concentrated intentionally for the purpose of doing work tend to be ions. H^+ and Na^+ are the usual suspects. Cells are quite adept at harnessing the potential energy stored in the concentration gradients of ions to do work, including making ATP.

Consider this analogy – you are one of 150 people who are packed tightly into a very small room. There is one doorway out of this incredibly crowded room, but it is blocked by a bouncer who demands $5 to exit. After ten minutes in uncomfortably tight quarters, you decide it's worth it to pay $5 to leave. In fact, let's say 50 people are willing to pay $5 to get out of the cramped room. The bouncer collects a total of $250 as 50 people leave. Now there's only 100 people left and with a little more breathing room, those 100 remaining people think $5 to exit is too much. The bouncer lowers the cost of exit to $3 each. 20 more people are now willing to leave at $3 each. *Paying money to leave* is analogous to the *release of energy* as a concentration gradient diffuses. Note that the more crowded the room, the more energy ($) was paid out as the concentration gradient diffused. As the room becomes less crowded, less energy is released.

Problem 11

Each image shows a concentration gradient of protons (H^+ ions) separated by a membrane. Assume the only way the protons can move into the lower compartment is through the membrane protein (purple), which is **ATP synthase**. *Which scenario releases the most energy as H^+ moves from the top compartment to the bottom compartment?*

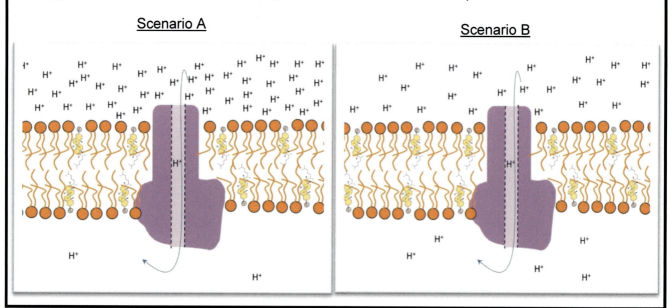

If ions of the same charge are crowded into a small compartment, they are especially keen to leave. This is because like charges repel each other, just like the third phosphate group is eager to pop off ATP because of its proximity to other negative charges. Thus, a concentration gradient of same-charged ions has more potential energy than a similar concentration gradient of an uncharged molecule (such as water).

We are coming full circle to our final topic of aerobic respiration. When biology students are first introduced to the process of aerobic respiration, the typical response is something like the following cartoon (**Figure 5**). That's unfortunate, because aerobic respiration is truly awesome. And its awesomeness should inspire great interest even in the most tepid student. The lack of student engagement (putting it mildly) is because students get bogged down in the complex details. Students fail to "see the forest for the trees", so to speak.

Figure 5. Le Rage cartoon demonstrating the common sentiment most students have about cellular respiration when they first learn it. Comic by Nalini Biggs.

The rest of this chapter outlines the "forest" of aerobic respiration with a strong emphasis on the big picture. After all, if you lose sight of the big picture, then the details don't make any sense anyhow. Let's start with the punchline. *What's the point of aerobic respiration?*

The purpose of aerobic respiration is to attach the third phosphate group back onto ADP to create ATP.

Anytime you attach a phosphate group onto something, it's called **phosphorylation**. Thus, the endgame of aerobic respiration is to *phosphorylate ADP* (reference **Figure 4** as needed). As you've learned, pushing that third phosphate group onto ADP requires a large energy payment. In aerobic respiration, that energy payment comes from chemical energy, specifically the nonpolar covalent bonds in our fuels – sugars, amino acids and fatty acids. The more nonpolar covalent bonds in one of these fuel molecules, the more ATP can be made.[6]

On a side note......

Most adipose is white (or yellowish), but newborns have "brown fat" located along the back and shoulders. The name "brown" reflects the rich supply of blood vessels to the area, creating more of a brownish color. Typically, about half the chemical energy stored in a fat molecule is transformed into ATP energy. The purpose of brown fat is different - the energy stored in these fat molecules is released almost entirely as heat (rather than recaptured as ATP chemical energy). When brown fat is metabolized, the goal is to generate heat, which is then distributed through the body by blood. The brown fat is like a molecular furnace that keeps a newborn warm.

The basic process is to *oxidize* sugars, amino acids or fatty acids by swapping out the high energy nonpolar covalent bonds for low energy polar covalent bonds. The term *oxidize* means a shifting in % of electron ownership. Remember that the high-energy nonpolar bonds are C-C and C-H. From carbon's perspective, this is a *reduced* state. When the organic molecule is oxidized, these nonpolar bonds are replaced with low energy polar covalent bonds of C-O and O-H. Carbon has been oxidized as its new partner (oxygen) hogs the electrons. The entire process is classified as an oxidation-reduction reaction, catabolic, and highly exergonic.

In a general biology course, aerobic respiration is introduced with glucose as the starting fuel. This is intentional; glucose is the most common starting reactant in living organisms (though certainly not the only fuel). We'll adhere to the trend and use glucose as our starting reactant too.

The process of aerobic respiration gets broken down into "stages". In a basic biology textbook, the stages are as follows:

    (1) Glycolysis

    (2) _____ cycle that goes by three different names: *Citric acid*, *Krebs*, or *TCA*

    (3) Oxidative phosphorylation

Additionally, there is a linkage step between stages 1 and 2 that may be taught as it's own distinct phase. And it's somewhere around here that students' eyes start glazing over as students become lost quickly in the details. Don't. It's not that hard, really. Breathe. Come on. Breathe. And try to focus, because I'm about to simplify if for you.

---

[6] This is why fatty acids, which are almost entirely nonpolar covalent bonds, yield more energy than carbohydrates or proteins.

Stages 1 and 2 (*glycolysis* and *goes-by-three-names cycle but is most commonly known as the citric acid cycle*) is the collection of over 20 chemical reactions that do this - **break the nonpolar covalent bonds of glucose (high energy) and create six low energy CO_2 molecules as a byproduct**. Did you get that? If not, watch this quick video demonstrating it for you.

The metabolic pathways of glycolysis and the citric acid cycle release energy. Some of that energy directly pays for the endergonic reaction of ADP + P → ATP. But not as much as you'd think. In fact, only about 4 ATP molecules are produced from one glucose molecule during the first two stages.

$C_6H_{12}O_6 + 6\ O_2 \rightarrow \rightarrow 6\ CO_2 + 6\ H_2O +$ 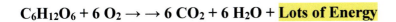 Lots of Energy

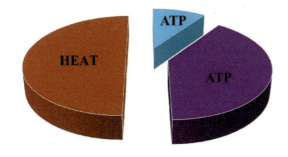

■ ATP directly from glycolysis and citric acid cycle
■ ATP from oxidative phosphorylation

Figure 6. A pie chart showing the relative distribution of energy that is released during aerobic respiration.

Figure 6 shows where the energy released from aerobic respiration goes. About half is recaptured in the form of ATP; ATP is the usable energy "currency" to a cell. The other half of the energy is released as heat. Aerobic respiration should yield between 30 – 36 ATP per glucose molecule. Yet only 4 ATP are gained from the first two stages. Most of the ATP comes from the final stage – oxidative phosphorylation.

Here's how oxidative phosphorylation works.

By the end of the citric acid cycle, glucose has been completely wittled away into 6 CO_2 molecules; CO_2 leaves the cell as a waste product. But remember that glucose is $C_6H_{12}O_6$. Where did the 12 hydrogen atoms go? During glycolysis and the citric acid cycle, the hydrogen atoms are picked off glucose and transferred to another molecule temporarily; this molecule acts like a *hydrogen* "taxicab".

These "taxicab" molecules[7] have empty spots that can hold one or two protons (H^+) and one or two electrons (**Figure 7**). The most common of these "taxicab" molecules is NAD, represented by the green car in **Figure 7**.

The chemical equation is written out like this: $NAD^+ + 1\ H^+ + 2\ e^- \rightarrow NADH$. Essentially, a "taxicab" molecule (NAD) picks up a hydrogen atom with one electron (that would be H^+ and e^- respectively) plus an extra electron. *Empty NAD is positively charged.* So, when NAD^+ picks up H^+ and two electrons, the final product is neutral and is represented as NADH (the final H indicating the *hydrogen* passenger). By the way, the passengers are an *all-or-none* deal. Either NAD is empty, or it has all three passengers.

NAD is the most common taxicab molecule, but it's not the only one used in aerobic respiration. The other one is FAD, which functions in a similar manner.

---

[7] These molecules are actually a modified nucleotide, but the names are so long, the initials are used instead. For example, the real name of NAD is Nicotinamide Adenine Dinucleotide (NAD).

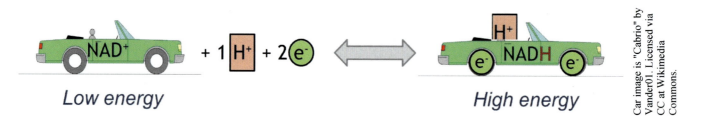

Figure 7. Chemical equation for the coenzyme NAD. NAD⁺ picks up its "passengers" to become NADH.

Problem 12

The following reactions are *redox* reactions involving two "taxicab" molecules used in aerobic respiration. For each reaction, circle the oxidized state and put a box around the reduced state.

$$NAD^+ + 1\ H^+ + 2\ e^- \rightarrow NADH$$

$$FAD + 2\ H^+ + 2\ e^- \rightarrow FADH_2$$

Problem 13

Look again at Figure 7. Which form of NAD has more chemical energy?

    a) the oxidized form   b) the reduced form

Problem 14

Are the reactions in Problem 12 *exergonic* or *endergonic*? _____

Review your answers to Problems 12 13, and 14 with the answer key at the end of this chapter before proceeding.

Time for a review. During the catabolism of glucose, hydrogen atoms are transferred to various NAD⁺ and FAD "taxicabs" during the first two stages. So, by the end of the citric acid cycle, here's what we have: 4 ATP molecules, 10 NADH molecules, 2 $FADH_2$ molecules, and 6 CO_2 molecules (which leave the cell as a waste product). The reduced forms of NAD and FAD (NADH and $FADH_2$ respectively) have more chemical energy than the oxidized form, so <u>NADH and $FADH_2$ represent high energy molecules.</u>

In summary, much of the chemical energy released from glucose was **not** directly transferred to the high energy molecule of ATP, but rather transferred to the "taxicabs" to create the high energy molecules of $FADH_2$ and NADH!

In my class, I sometimes equate NADH and FADH₂ molecules with casino chips. Casino chips *represent* money, but they have to be exchanged for cash before the currency can be spent in a store (**Figure 8**). Likewise, NADH and FADH₂ have energy, but it's not spendable in the cell. Thus, the final stage of aerobic respiration is all about exchanging NADH and FADH₂ for usable ATP energy. This is the purpose of oxidative phosphorylation.

Figure 8. In oxidative phosphorylation, NADH and FADH₂ are "cashed in" for 3 ATP and 2 ATP respectively.

Let's start with the name *oxidative phosphorylation*. From your knowledge of redox reactions, you may infer that the word "oxidative" means *to oxidate*, or remove electrons from, something. What is oxidated in this process? NADH and FADH₂ of course! Remember our "taxicab" molecules were just temporary holders of those electrons and hydrogen(s). Now it's time to drop them off. The word *phosphorylate* means to add a phosphate group. What is getting phosphorylated? ADP! So, *oxidative phosphorylation* is an energy transformation. The energy released from *oxidizing* NADH and FADH₂ powers the the endergonic phosphorylation of ADP to create ATP (**Figure 9**). (Better) living through chemistry.

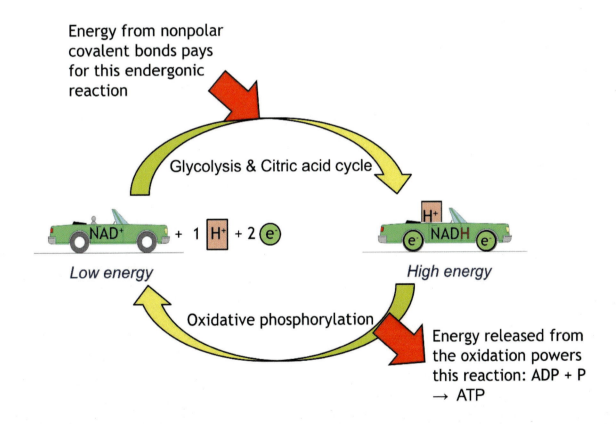

Figure 9. The "taxicabs" rotate between a state of empty (NAD⁺, FAD) and full (NADH, FADH₂). The empty state is the oxidized form; the full state is the reduced form. Because reducing (or adding passengers to) the taxicabs is endergonic, energy payment is required for the reaction to occur. Energy payment occurs during glycolysis and the citric acid cycle by breaking nonpolar covalent bonds from a fuel molecule. The NADH and FADH₂ then shuttle over to the electron transport chain and drop off the electrons. The oxidation (removal of electrons) of NADH and FADH₂ releases energy, which is converted to the chemical energy of ATP.

NADH and FADH$_2$ drop off their "passengers" (hydrogen atoms and electrons) at a membrane protein that is part of the **electron transport chain**. The electron transport chain is exactly what the name suggests – it is a *chain* of membrane proteins that passes electrons from one protein to another. Sounds exciting, doesn't it? No, I'm not being sarcastic. It really is pretty awesome. Read on.

In a human cell, the electron transport chain is embedded in the inner membrane of the mitochondria. Watch this video to see how NADH and FADH$_2$ become oxidized by the electron transport chain.

In summary, the oxidation of NADH and FADH$_2$ create a concentration gradient of H$^+$ in one mitochondrial compartment. Remember that this concentration gradient is potential energy that can be used to do work. When H$^+$ diffuses back into the other mitochondrial compartment, it does so ONLY through a special membrane protein called ATP synthase. The name of the membrane protein describes what it does – *synthesize ATP*. ATP synthase converts the potential energy of an H$^+$ concentration gradient into the potential energy of ATP.

Now that you've learned the summary of oxidative phosphorylation, examine **Figure 10**, which puts the pieces together.

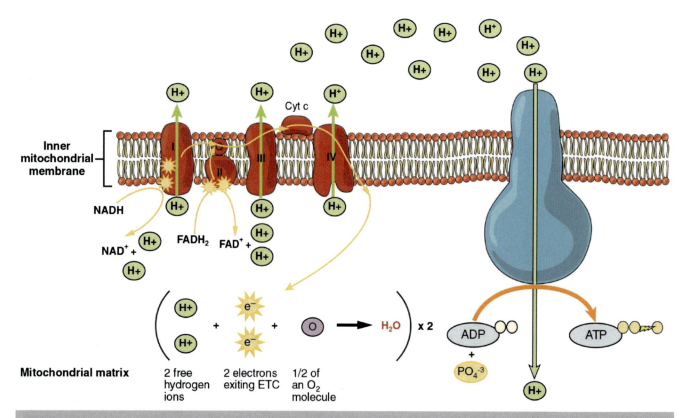

Figure 10. The summary of oxidative phosphorylation. Image from OpenStax College licensed by Creative Commons via Wikimedia Commons

The final piece of oxidative phosphorylation is O$_2$, which is the *final electron acceptor* in the electron transport chain. What does this mean? Essentially, O$_2$ takes the electrons from the last protein in the chain and keeps them. Hence, the *final* acceptor of electrons. Once O$_2$ has four electrons, it combines with free H$^+$ ions (that have just diffused through ATP synthase) to make water. Look at **Figure 10** again for reference.

Congratulations if you've made it to this point! That material is challenging and you should be proud of yourself for getting through it. There's just a few more problems left for you to do. Please don't hesitate to re-read parts of this chapter to reinforce your knowledge. Almost no one fully understands it the first time through.

Problem 15

In Figure 10, label the Electron Transport Chain and ATP Synthase.

Problem 16

In this image of a mitochondrion, identify where the H⁺ ions are concentrated.

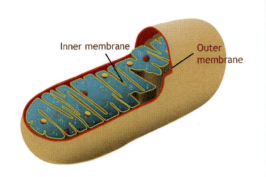

Problem 17

The summary of aerobic respiration is written as $C_6H_{12}O_6 + 6O_2 \rightarrow 6 H_2O + 6 CO_2$.

1. At what stage(s) is O_2 consumed?
   a) Glycolysis and/or citric acid cycle b) oxidative phosphorylation

2. At what stage(s) is CO_2 produced?
   a) Glycolysis and/or citric acid cycle b) oxidative phosphorylation

3. At what stage(s) is water produced?
   b) Glycolysis and/or citric acid cycle b) oxidative phosphorylation

Problem 18

Glycolysis and the citric acid cycle produce 10 NADH and 2 FADH₂ per glucose. **Figure 8** shows that each NADH is worth ~ 3 ATP molecules and each FADH₂ is worth ~ 2 ATP molecules. If 10 NADH and 2 FADH₂ molecules are "cashed in" during oxidative phosphorylation, how many ATP are generated during oxidative phosphorylation[8] per glucose?

---

[8] The number that you calculate is an upper limit. The actual ATP yield is between 10 – 20% less than the upper limit for reasons beyond the scope of this tutorial.

Chapter 10: Answer key

1. Kinetic – light, heat, motion; Potential – chemical energy (specifically in covalent bonds), concentration gradient

2. Only the third graph satisfies both laws of thermodynamics.

    Explanation: The second graph does not satisfy the second law of thermodynamics (adding heat to a system and converting it to chemical energy increases order and therefore decreases entropy). The fourth graph does not satisfy the first law of thermodynamics (energy must be conserved).

3. 1, 3, and 4 are spontaneous because there are more nonpolar covalent bonds in the reactants than the products; 1, 3, and 4 are energetically favorable.

4. Fatty acid. The rule is the more nonpolar covalent bonds, the more energy. Glucose and the polar amino acid each have quite a few polar covalent bonds while the fatty acid has very few. Thus a gram of fatty acids will have more chemical energy than either a gram of amino acids or a gram of glucose.

5. 1, 3, 4 are exergonic. 2 is endergonic.

6. Reactions 2, 3 and 6 are catabolic. Reactions 4 and 5 are anabolic.

7. Ca^+ loses another electron to become Ca^{2+}. Calcium has been oxidized. Each chlorine atom gains an electron to become Cl^-. Chlorine has been reduced.

8.

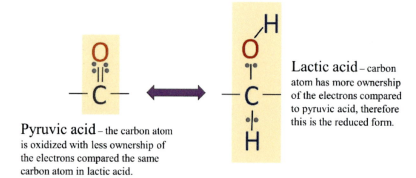

9. b

    Explanation: Going from lactic acid to pyruvic acid is slightly energetically favorable as in this direction, the redox reaction moves electrons towards the atom with the highest affinity for them – oxygen. Anytime the redox reaction moves electrons to an atom or molecule that wants it more, the reaction is spontaneous, or energetically favorable.

10. 1) c   2) d

11. Scenario A

12. $NAD^+ + 1\ H^+ + 2\ e^- \rightarrow NADH$ $FAD + 2\ H^+ + 2\ e^- \rightarrow FADH_2$

13. b) the reduced form

14. endergonic

15.

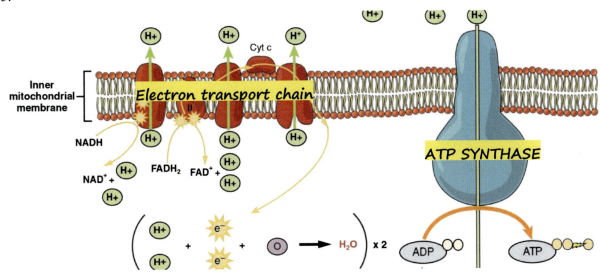

16. H⁺ ions (also known as protons) are concentrated in the intermembrane space (colored in yellow).

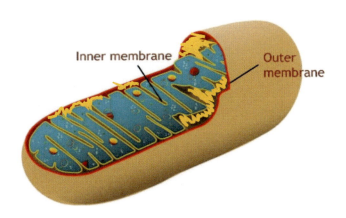

17. 1) b   2) a   3) b

18. 10 NADH x 3 ATP/NADH = 30 ATP

   2 FADH₂ x 2 ATP/FADH₂ = 4 ATP

   Total ATP generated in *oxidative phosphorylation* = 30 + 4 = 34

Chapter 11: Molecular transport

Molecular transport is the "heartbeat" of physiology.

Molecules, ions and other small substances are in constant motion; the vibrating motion moves particles in random directions. The result of these random motions is diffusion. **Diffusion** is when particles do not maintain a sustained concentrated presence in any one area. Instead, the particles naturally spread out within their confines until they are no longer concentrated. For example, when a drop of red dye is added to a cup of water, the dye molecules start out extremely concentrated but eventually spread out until no one concentrated area exists (**Figure 1**). Likewise, think of cologne or perfume. The smell is initially concentrated in the first spritz. After an hour or so, the entire room smells faintly of the odor, but no one area is particularly pungent.

Figure 1. (Left) Red food coloring is initially concentrated. (Right) After 30 minutes, the dye molecules are dispersed evenly in the water, and the dye has reached diffusion equilibrium.

In order for diffusion to occur, a **concentration gradient** must first exist. A concentration gradient is where particles are more concentrated in one area than another (e.g. the initial drop of food coloring). Diffusion is the process of particles becoming less concentrated and the appropriate terminology is say particles move *down* a concentration gradient. When particles are no longer concentrated in one any area, we say that a state of **diffusion equilibrium** has been reached. Use your smart phone or tablet to access this QR code for a short video of small children exhibiting diffusion. Like the children in the video, diffusion is a natural consequence of motion (it's not as if molecules purposefully try to spread out). Therefore, diffusion is considered a **passive process** because the cell does not expend energy to make diffusion happen. Diffusion is pertinent to cellular processes as we shall see in this chapter.

Problem 1

A drop of concentrated green particles are placed in a box. Which box best represents **diffusion equilibrium**? (*Hint: only one answer is correct*).

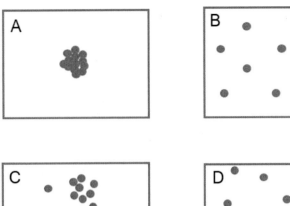

There are several factors that affect **diffusion rates**, or how quickly a substance reaches diffusion equilibrium. These three factors are important in physiology: (1) the size of the particle diffusing, (2) the temperature of the environment, and (3) the degree to which the particles are concentrated.

Because diffusion is a consequence of random motion, it makes sense that the speed at which particles move affects the rate of diffusion. So what influences the speed at which particles move? Temperature and size. Temperature is a measurement of kinetic motion; higher temperatures reflect faster moving molecules. Particle size matters too. Lightweight molecules and ions zip about quickly. Heavier molecules or particles lumber around more slowly. Consequently, the diffusion rate of larger particles is slower too. The third factor is the degree to which a substance is concentrated. Please use your smartphone or tablet to access this QR code which describes how these three factors affect diffusion rates.

Problem 2

For each *pair* of underlined scenarios, (circle) the scenario that exhibits a **faster** rate of diffusion.

1. Diffusion occurring at 92°F *or* diffusion occurring at 98°F.

2. Diffusion occurring with H⁺ ions *or* diffusion occurring with glucose molecules.

3. Diffusion occurring where one area is 50x more concentrated than the other area, *or* diffusion occurring where one area is 2x more concentrated than the other area.

Now that we understand the natural trend of particles to spread out while in motion, consider **Figure 2**. Is the plasma membrane a barrier to diffusion of the green particles? The answer is (drumroll)..... *it depends*.

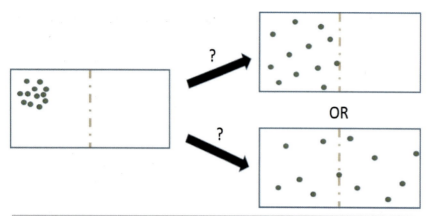

What does it depend on? That's a big question with a complicated answer. Nonetheless, we will answer it, for understanding if and how substances move in and out of the cell is hugely important in physiology. Note that it is the job of the plasma membrane to regulate and control such molecular transport[1].

Figure 2. Left image - green particles exhibit a concentration gradient that will be alleviated as the particles begin to move. Right image - two possible outcomes of diffusion, depending on whether or not the membrane (orange dashed line) is a barrier to the green particles.

First, let's identify the three avenues by which a molecule *could* cross the plasma membrane (**Figure 3**).

A molecule could traverse the plasma membrane through …

(1) gaps in the lipid bilayer,

(2) a transmembrane protein designed for molecular transport, or

(3) vesicles that interact with the plasma membrane.

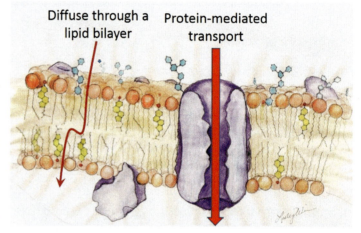

Figure 3. Two of the three mechanisms for molecular transport are shown. Few molecules can diffuse through the temporary gaps created in the bilayer as the lipids move. More molecules require protein assistance in what is known as **protein-mediated transport**.

Most molecules cross by only one of these mechanisms. Glucose, for example, crosses the plasma membrane only through transmembrane proteins.

We will explore all of these pathways in detail, starting with identifying which molecules pass through gaps in the lipid bilayer.

In **Chapter 9** we learned that the phospholipids in the plasma membrane flex their tails and move laterally, creating temporary openings that allow some small molecules to tentatively enter the bilayer. Once a molecule "puts a toe" into the bilayer, it is confronted with the very hydrophobic lipid core (**Figure 4**).

---

[1] In reference to transport, the term "molecule" is applied loosely. Ions (which can be single atoms rather than molecules) are included in the discussion of "molecules" throughout this chapter.

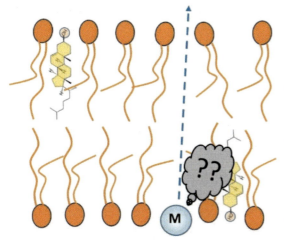

Figure 4. A molecule (M) seizes an opportunity to wade into the bilayer as two phospholipid heads temporarily move apart from each other. Can M move across the bilayer?

Does the molecule cross the hydrophobic core and get to the other side? That depends on three factors: (1) the size of the gap created by the moving phospholipids, (2) the size of the molecule attempting to cross, and (3) how charged the molecule is. Let's look at these three factors independently.

The size and frequency of the gaps created in the lipid bilayer depends on *how fluid* the bilayer is. We learned in Chapter 9 that the primary determinant of fluidity in a human membrane is the amount of cholesterol in it. Very permeable membranes have little cholesterol and therefore make lots of gaps (and big ones at that). The more gaps created and the larger the gaps, the more likely a molecule can pass through a temporary gap in the lipid bilayer.

The impact of molecular size is straightforward. Big molecules have trouble finding an opening large enough to accommodate their passage through the bilayer. Smaller molecules have more opportunities. They can take advantage of small or medium-sized gaps, which are far more plentiful. Smaller molecules also move faster and can quickly dart across that temporary opening before it closes. Thus, the smaller the molecule, the better its chances of crossing a lipid bilayer. It turns out that all molecules capable of diffusing through a lipid bilayer are smaller than a phospholipid; most are much smaller.

The third factor is how charged the molecule is. Here's the set up. The phospholipid tails and cholesterol rings of the lipid bilayer core consist entirely of carbon and hydrogen atoms. Therefore, the membrane core is as nonpolar as it gets. We generally refer to these nonpolar lipids as *hydrophobic* because they are repelled by water. Let's change the perspective. We are interested in how our potential travelers (molecules attempting to cross a lipid bilayer) respond to the lipids, rather than water. We use the terms **lipophilic** and **lipophobic** to describe a molecule's reaction to the lipid bilayer core. Now, it's time to introduce the lipid *love-o-meter* ♥. Place your smartphone or tablet over the following QR code for the explanation!

Problem 3 *Challenge!*

In the following list, (circle) any solute that is capable of diffusing through a phospholipid bilayer. (Hint: 3 are correct).

Na^+

Glucose

Any amino acid

A large protein

CO_2

Fat

O_2

Cl^-

A fatty acid (with no charge)

Compare your answer to Problem 3 with the answer key at the end of this chapter before proceeding.

If a molecule is capable of diffusing through a phospholipid bilayer, then the plasma membrane is either a poor or nonexistent barrier to that particular molecule. These molecules - **O_2, CO_2, fatty acids** and even many **steroids** - are "ghosts" that recognize no boundaries (much like a ghost would not be contained by the walls of a room). In general, these molecules cannot be concentrated within a cell or purposefully excluded from a cell. This has significant consequences to our biology, as the following example will demonstrate.

The reactants for aerobic respiration are glucose ($C_6H_{12}O_6$) and O_2; both glucose and O_2 are required to make ATP. As you've just learned, a phospholipid bilayer is an effective barrier for glucose. Therefore the cell can store glucose. However, O_2 freely crosses the phospholipid bilayer. If a cell tried to concentrate O_2 like it stores glucose, the O_2 molecules would just diffuse back through the plasma membrane until the intracellular and extracellular concentrations of O_2 were equal.

Let's say a cell needs 100 glucose and 600 O_2 molecules to make 3400 ATP molecules. The glucose molecules can be stored in the cell, but the O_2 cannot. Will there be enough O_2 to make 3400 ATP molecules? That depends on the O_2 concentration in the blood. The O_2 concentration in the cell will mirror the O_2 concentration of the blood (extracellular region). *This is why it's necessary to breathe constantly, but not eat constantly.* The respiratory system must continually load up your blood with as much O_2 as possible to create favorable gradients to maximize the O_2 diffusion into your cells (**Figure 5**). Hypothetically, imagine if O_2 could be stored like glucose! You could hyperventilate for 5 minutes, capture all the O_2 you need and store it, and then you wouldn't need to breathe again for a few hours[2]. Breathing patterns would be just like eating patterns!

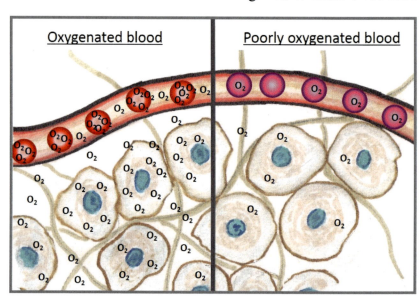

Figure 5. The concentration of O_2 in the cells reflects the concentration of O_2 in the blood. Continuous breathing aims to keep the blood oxygen concentration high.

When determining whether molecules *can* or *cannot* diffuse across the lipid bilayer, very small polar molecules represent a "grey area" of sorts. It turns out that the smallest of these – water – seems to pass through rather readily, assuming the membrane has a low cholesterol content. This usually surprises students as the notion goes against the very grain of the term *hydrophobic*. But remember, on the *Lipid love-o-meter*, water is not public enemy #1. Ions are. So while polar molecules are ill-tolerated, water molecules are so small, and in such abundance, that they do manage to zip across the temporary breaches

---

[2] The sperm whale (a marine mammal) can hold its breath for 1 – 2 hours while diving. How? There are many adaptations that assist in the prolonged breath-holding, but a notable one is this: 20% of their body mass is blood; the extra blood functions as an O_2 reservoir during extended dives. In contrast, blood makes up only 6-7% of a human's body mass.

in the phospholipid bilayer in fair numbers. In most cases, water molecules should be thought of like "ghosts". We'll return to the diffusion of water shortly.

Problem 4

List five molecules that readily cross a plasma membrane through temporary gaps in the lipid bilayer.

Molecules *other* than those listed in Problem 4 <u>must</u> be transported across the plasma membrane by one of the other two mechanisms - a transmembrane protein or through vesicular transport.

Smaller molecules (e.g. monomers, small polar molecules and ions) usually travel in and out of the cell by transmembrane proteins[3]. This transport is called **protein-mediated transport** because it requires the assistance of a membrane protein to cross the plasma membrane. Membrane transporters come in two basic designs – **channel** and **carrier** (**Figure 6**). Put your smartphone or tablet over the following QR code to watch a simple animation depicting the differences between each type of membrane protein.

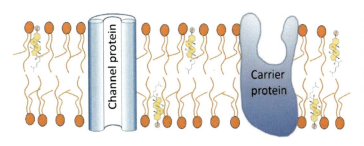

Figure 6. The anatomies of a channel and carrier protein.

Problem 5

What type of membrane transporter – *channel* or *carrier* – is shown here?

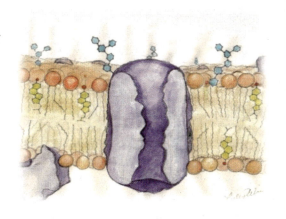

---

[3] Membrane proteins involved in transport go by several monikers – *membrane transporters*, *protein transporters*, or even just *channels* or *carriers*. You may also see the phrase *protein-mediated transport* used to reference membrane transporters.

Each membrane transporter is highly specific. There is no *one-size-transports-all* membrane transporter. To transport Na⁺, the cell would insert a transmembrane protein specifically configured to allow passage of Na⁺ only. Transport of Cl⁻ would require a different protein altogether. Therefore, Cl⁻ wouldn't travel through a *Na⁺ channel* and vice versa. Each type of transporter has a name, and it's usually based on what solute they transport (e.g. **GLUT4** transporter transports **glu**cose; Na⁺ channel allows for sodium passage).

The third route across the plasma membrane is to travel using vesicles that integrate with the plasma membrane. **Endocytosis** is the process of pinching off a vesicle from the plasma membrane to bring stuff into the cell (**Figure 7**). **Exocytosis** is the reverse, whereas a vesicle carrying molecules from the Golgi body fuses with the plasma membrane. As part of the fusion, the molecules that were inside the vesicle are pushed to the outside of the cell. Notably, moving vesicles along the cytoskeleton tubules requires ATP catabolism. Thus, endocytosis and exocytosis are energetically-expensive processes.

Knowing the meaning of the prefixes will help you distinguish between these terms. *Endo-* means internal or within. Endocytosis brings substances into the cell. *Exo -* means outward or outside, like *exit* or *export*. For example, if a vesicle carrying a large batch of hormones from the Golgi body fuses with the plasma membrane to secrete the molecules outside of the cell, that's exocytosis.

Figure 7. Three types of endocytosis. In each case, notice that the plasma membrane indents into a bubble (vesicle) carrying stuff into the cell. Phagocytosis is important in microbiology, whereas receptor-mediated endocytosis is more significant in physiology.

When I think of endocytosis, I envision the process like blowing bubbles. A giant soap film in a bubble wand acts like the plasma membrane. As a person blows gently into the soap film, the film bulges outward until the bulge breaks away as a spherical bubble (vesicle). The soap film reforms in the bubble wand, much like the plasma membrane reseals itself after a vesicle has pinched off. Exocytosis is like watching someone blow bubbles in rewind.

Photo by Sagar Joshi Licensed by CreativeCommons Wikimedia 3.0

Problem 6

These pictures represent a sequence of events numbered sequentially from 1 through 4. ECF = extracellular fluid; ICF = intracellular fluid. Select the correct word to finish each sentence.

1. This series of pictures is showing _____.
   a) endocytosis
   b) exocytosis

2. The process shown allows the cell to _____ molecules.
   a) secrete
   b) ingest

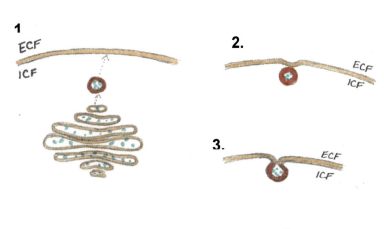

Why would a cell use vesicles for transport (i.e. exocytosis or endocytosis) rather than membrane transporters? There are several scenarios when vesicles are the better or only option. A cell will use vesicles for transport when the molecules are too large to move with membrane transporters. Indeed, even carrier proteins have a limit to the size of solute they can transport. Peptides and proteins are simply too large to move with a carrier protein and so must be transported with vesicles. Sometimes a molecule could be moved via a membrane transporter, but because so many need to be transported at once, a vesicle is just more efficient. Think of it like using a large bus to shuttle a group of people to the same destination rather than each person driving their own car.

Problem 7

List two scenarios where a cell would use vesicle transport (endocytosis or exocytosis) rather than membrane transporters to transport a substance.

1.

2.

Problem 8

Define the following acronyms.

ECF -   ICF -

Let's summarize what we've learned about crossing the plasma membrane into three *rules of thumb*.

 (1) Small lipids (e.g. steroids, fatty acids), O_2 and CO_2, and in most cases H_2O diffuse freely through the phospholipid bilayer. The plasma membrane is not a boundary to these "ghosts", so sustaining any sort of concentration gradient of these molecules is near impossible.

 (2) Ions, small polar molecules and monomers use membrane transporters (transmembrane proteins embedded in the plasma membrane that perform transport).

(3) Larger molecules such as peptides and proteins will use vesicles for movement in and out of the cell. Vesicles may also be used for smaller molecules that are transported in bulk.

As the solvent of all living beings, the passage of water through the plasma membrane deserves special attention. My rule of thumb states that water freely diffuses through a phospholipid bilayer "in most cases". It's important to define what those cases are. The first thing to consider is the percentage of cholesterol in the plasma membrane, which varies from cell to cell. Membranes with a 1:1 ratio of cholesterol to phospholipids are relatively impermeable to water. With just as many cholesterol molecules as phospholipids, the membrane's permeability to small molecules of any sort, including water, is greatly reduced. Conversely, low-cholesterol membranes are more fluid and therefore more permeable. If a cell wants water to move freely in and out of the cell, the plasma membrane will have a low cholesterol content. It's worth mentioning that even with low-cholesterol membranes, water diffusion through the bilayer is still slow; water molecules must bide their time waiting for temporary gaps to manifest in the lipid bilayer.

A cell can speed up water diffusion by inserting a channel protein specific for water. This appropriately named **aquaporin** (*pore* for *water*) provides safe passage for water molecules to quickly diffuse in or out of the cell. The more aquaporins in the plasma membrane, the more permeable the membrane is to water, and the faster net water movement in or out of a cell occurs. In summary, both cholesterol content and aquaporin prevalence determine the degree of permeability to water molecules. The need for a plasma membrane to vary its permeability to water will become evident when learning about kidney function in Human Physiology.

Problem 9

This plasma membrane is very permeable to water. List two ways a cell can alter this plasma membrane to make it LESS permeable to water.

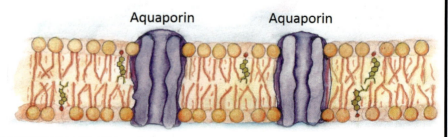

Like all diffusion processes, water molecules want to move <u>away</u> from areas of high water density. When biological membranes are permeable to water, water molecules will diffuse until the concentration of water molecules in the cell is the same as the water concentration outside of the cell.

When water diffuses across a membrane (e.g. the plasma membrane), the process gets a special name – **osmosis**. Making accurate predictions about whether or not osmosis will occur requires a solid understanding of **concentration,** specifically of water. In biology, *water concentration* is almost never stated; concentrations instead refer to a *solute* dissolved in water (recall, water is the solvent).

The first thing to understand is that the concentration of any solute depends on the <u>amount of the solute</u> AND the <u>volume of water</u>. Place your smartphone over the following QR code to watch a demonstration of how both factors affect concentration.

Solute concentrations can be represented three different ways. For example, a 2% glucose solution is also 20 g(rams) glucose/L(iter) of water, which is also a 0.111 M(olar) glucose solution. Thus, the same solution can be labeled by

(1) percentage (e.g. 2% glucose),
(2) weight per volume (e.g. 20 g glucose/L of water), or
(3) molarity (e.g. 0.111 M).

In a Human Physiology course, you may learn how to convert between the three representations. For now, understand that the higher the number, the more concentrated the solute is. For example, a 4% glucose solution is twice as concentrated as a 2% glucose solution. A 0.55 M solution of glucose is five times as concentrated in glucose as a 0.11 M solution. Try the following problem.

Problem 10

For each pair of solutions, (circle) the solution with a higher *solute* concentration. As an example, the first one has been done for you.

        5% urea solution   or   (10% NaCl solution)

    0.1 M glucose + 0.1 M urea solution   or   0.15 M glucose solution

20g/L glucose solution   or   10 g/L glucose + 10g/L maltose + 15 g/L KCl (salt) solution

---

We use the solute concentration to deduce the water concentration. If I have a 2% glucose solution, that means that 2% of the solution is glucose molecules. Because the total is 100%, *the other 98% of the solution must be water molecules.* So, a 2% glucose solution is also a 98% water solution. Now consider a 10% glucose solution. A 10% glucose solution is a 90% water solution. Here's the rule of thumb.

The *higher* the concentration of solute, the *lower* the concentration of water in an aqueous solution. The converse is also true.

Water concentration is easiest to calculate in the *percent* format, but you can apply the same rule of thumb to molarity and weight/volume descriptors of concentration too. Try this next problem.

Problem 11

For each pair of solutions, circle the solution with the *highest* concentration of water.

5% urea solution   or   10% NaCl solution

0.1 M glucose + 0.1 M urea solution   or   0.15 M glucose solution

20g/L glucose solution   or   10 g/L glucose + 10g/L maltose + 15 g/L KCl (salt) solution

---

When two solutions are separated by a membrane permeable to water, water molecules diffuse across that membrane towards the solution with a *lower concentration of water* (i.e. down water's concentration gradient). Invoke the rule of thumb above. This is the same as saying water molecules move towards the compartment *with more solutes*. Because concentrations are listed with respect to solutes (rather than water), it is easier to remember that in osmosis, *net water movement is into the compartment with a higher concentration of solutes.*

Figure 8 demonstrates osmosis through a membrane. In this image, a glass U tube is separated into two halves by a lipid bilayer. For a more thorough explanation of **Figure 8**, and why osmosis occurred but not salt diffusion, access the following QR code with your smartphone or tablet.

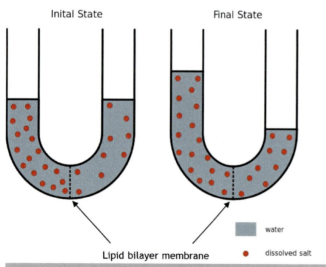

Figure 8. Osmosis occurs across a lipid bilayer when a concentration gradient exists and the membrane is impermeable to the solute.

Image modified from "Osmose en" by © Hans Hillewaert. Licensed under CC BY-SA 3.0 via Wikimedia Commons

Let's start applying this knowledge. In Problem 12, you will encounter a series of hypothetical "cells" that have just been dropped into a solution of extracellular fluid. Your job is to predict which substances will diffuse (and if so, in what direction) by applying your knowledge of diffusion, osmosis, and membrane permeability. The best way to introduce you to the flavor of these problems is to walk you through two examples. Please get out your smart phone or tablet (yes, again!) and access the following QR code for step-by-step instructions on how to solve these problems.

Now that you've watched the tutorial, try Problem 12.

Problem 12

Each problem shows a "cell" that has just been dropped into a solution of ECF. Draw in the cell and ECF *after* all solute diffusion and/or osmosis has occurred. First predict solute movement, then predict water movement (if any). If osmosis occurs, draw your cell bigger or smaller to account for the water gain or loss (to the cell) respectively.

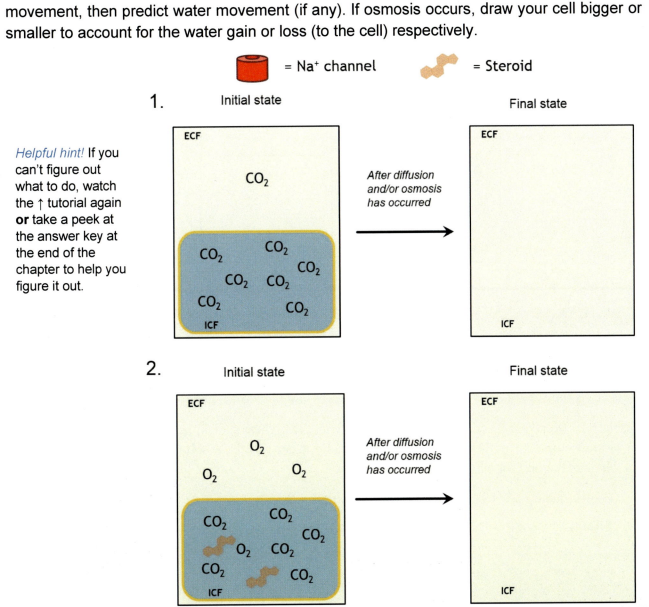

Helpful hint! If you can't figure out what to do, watch the ↑ tutorial again **or** take a peek at the answer key at the end of the chapter to help you figure it out.

149

Problem 12 continued

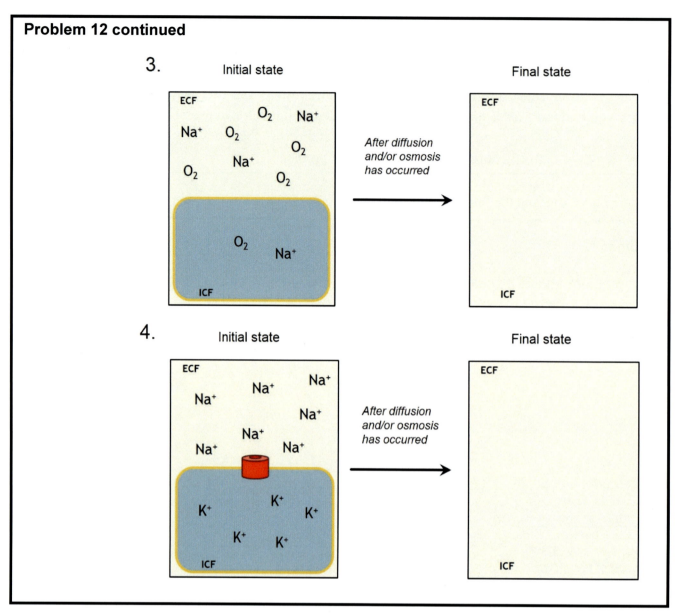

I strongly encourage you to compare your answers with the answer key before moving on.

In reality, hundreds of different solutes are dissolved in a human's intracellular fluid and extracellular fluid. Because water responds to the cumulative effects of all solutes, there is a shortcut method to quantify the total concentration of all solutes. It's called **osmolarity**, and it's a popular descriptor in Human Physiology. Osmolarity is the sum of each solute's concentration (in *molarity* form, hence os*molarity*).

In Problem 10, you were asked to select the solution with the highest concentration of solute: 0.1 M glucose + 0.1 M urea solution or 0.15 M glucose solution. Let's convert these solutes into their OsMolar (OsM) form, which is the unit for osmolarity.

➢ 0.1 M glucose + 0.1 M urea = 0.2 M total "stuff" = 0.2 OsM
➢ 0.15 M glucose = 0.15 M total "stuff" = 0.15 OsM.

As long as concentrations are already listed in molarity form (rather than % or weight/volume), calculating osmolarity is easy[4]. As you can see, 0.2 OsM is a greater solute concentration than 0.15 OsM. In case you are curious (you know you are!!), fluids in the human body (including both the ICF and ECF) are maintained around 0.29 OsM.

Osmolarity provides an easy method to compare the *total concentration of all solutes* between two solutions. To compare one solution's total solute concentration against another solution, the suffix *-osmotic* is paired with one of the following prefixes: *hyper-, iso-, hypo-*. In practice, we usually characterize the extracellular environment relative to the cell's fluid (ICF).

A **hyper**osmotic extracellular environment has **a higher** solute concentration than the cell.

An **iso**osmotic extracellular environment has **the same** solute concentration as the cell.

A **hypo**osmotic extracellular environment has **a lower** solute concentration than the cell.

For example, in Problem 12 – 1, the extracellular solution is **hypoosmotic** because there is only one solute in the ECF and seven solutes in the ICF. In Problem 12-3, the external solution has a higher concentration of solutes (eight solutes) relative to the cell's two, so the external solution is **hyperosmotic**. An important point: compare solute concentrations only in the *initial state* before any diffusion or osmosis occurs. This may seem counterintuitive until you realize that in the *final state*, all solutions will be isoosmotic to each other because of osmosis.

Problem 13

Complete each sentence with the correct word from this list: hyperosmotic, isoosmotic, hypoosmotic.

In Problem 12-2, the ECF is _____ relative to the ICF.

In Problem 12-4, The ECF is _____ relative to the ICF.

Osmosis can significantly affect cell volume, with potentially severe consequences. If the cell shrinks too much, chemical activities can slow down or stop. With extreme dehydration, the cell may senesce (stop working) or die. In microbiology, dehydrating bacteria cells with a hyperosmotic solution is actually an antimicrobial technique (a process designed to prevent microbial growth). In young plants, the cellular cytoplasm can shrink to a point of extreme wilting with the entire plant flopping over. Use your smartphone or tablet to access a video showing you the results of soaking a tomato plant in hyperosmotic salt water.

---

[4] Calculating osmolarity is strictly additive except when salts are involved. You'll probably learn about that exception in your Human Physiology course. That exception is beyond the scope of this tutorial.

What if water enters the cell? The cell begins to swell like an expanding water balloon. Bacteria, fungi and many protists have a cell wall that counteracts the **osmotic pressure** (the pressure that results when water is drawn into a solution with a higher solute concentration). The cell wall prevents the cell from getting too large. Human cells fare worse. Without a protective cell wall[5], the plasma membrane can be stretched to the point of snapping (think of overfilling a water balloon), which **lyses** (bursts) and kill the cell.

> *On a side note......*
>
> Rubbing salt onto the surface of a meat creates a hyperosmotic environment. Any opportunistic microbes looking to colonize the meat's surface are faced with cytoplasm shrinkage as a result of the hyperosmotic environment. Without sufficient cytoplasm volume, most cellular processes come to a screeching halt. Thus, the meat is "cured", as in "curing" your meat of any microbial spoilage. Creating a solute-enriched environment with either excess salt or sugar (e.g. jams) is an antimicrobial approach that has traditionally been used to preserve foods.

Problem 14 *Challenge!*

The top picture shows a typical plant cell from a leaf. The bottom picture is of a human red blood cell. **Draw** what both cells will look like after exposure to pure water. (*Hint:* assume each cell is packed full of various solutes that are not able to diffuse out of the cell).

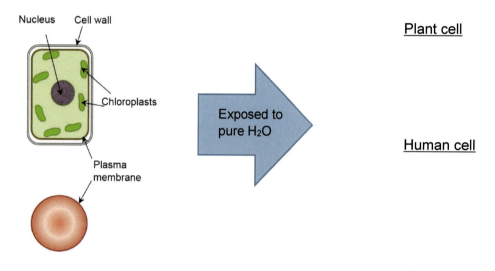

Plant cell

Human cell

Problem 15

Complete this sentence with the correct word.

Pure water is _____ compared to a cell.

    a) hyperosmotic   b) isoosmotic   c) hypoosmotic

---

[5] Like all animal cells, human cells do not have an external cell wall.

So far, we've discussed osmotic gradients (when one solution has a higher concentration of solutes than another) and osmolarity. **Tonicity** is a similar concept. Osmolarity simply quantifies the total concentration of solute, while tonicity is used to describe whether osmosis *will occur* between two solutions separated by a membrane. Confused? If yes, you are not alone. Most students treat *osmotic gradients* and *tonicity* as the same thing unless they are shown how the two terms are different. But they are different, as I shall demonstrate shortly.

For tonicity, we use the prefixes *hyper-*, *iso-*, and *hypo-* (recognize these?) applied to the suffix *–tonic* to describe how one solution affects the water volume of the solution on the other side of the membrane. In practice, tonicity usually describes how the external solution affects the volume of the cell's intracellular fluid. With that context, here are the definitions.

> A **hyper**tonic environment draws water out of the cell, causing the cell to shrink.

> An **iso**tonic environment does not change the volume of the cell.

> A **hypo**tonic environment causes water to enter the cell, and the cell swells.

So, a **hyper**tonic solutions causes a cell to shrink; a **hypo**tonic solution causes a cell to grow. How are you going to keep those two similar terms straight? One way I remember is it to write hyp**O**tonic with a big **O** in the word. The "big **O**" represents the *enlarged cell* that results from a hyp**O**tonic environment.

Let's illustrate the difference between tonicity and osmolarity with two examples. **Figure 9** shows hypothetical cells bathed in ECF. In **Figure 9A**, the ECF is hypoosmotic, but isotonic to the ICF. *Why?*

> ➢ There are fewer solutes in the ECF compared to the ICF, so by definition the ECF is hypoosmotic.

> ➢ But…. all solutes (O_2, CO_2, steroids) are permeable to the lipid bilayer so each solute achieves diffusion equilibrium. Because there is no lasting concentration gradient of solutes, no osmosis occurs. The ECF must be isotonic, having no lasting effect on the water volume of the cell.

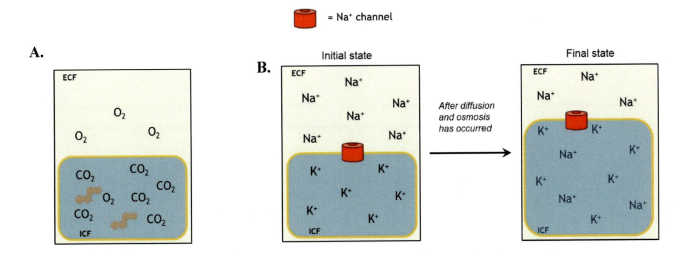

Figure 9. A. ECF is hypoosmotic but isotonic. B. ECF is isoosmotic but hypotonic.

Figure 9B shows the opposite trend. The ECF is isoosmotic to the ICF – the initial osmolarity of the ECF is the same as the ICF. However, Na^+ is allowed to diffuse through the channel protein. Therefore, some Na^+ ions enter the cell as Na^+ approaches its diffusion equilibrium. This increases the total solute concentration (osmolarity) in the cell. Water then moves into the cell (remember water moves into the compartment with more "stuff"), increasing the cell's volume. Hence, the ECF solution caused the cell to gain water. By definition, the ECF is hypotonic[6].

The examples presented in Figure 9 demonstrate that *osmolarity* and *tonicity* are not synonymous. However, there are plenty of cases where the osmolarity and tonicity share the same prefix. The tomato plant doused in salt water (QR code on page 151) is a great example. In that example, salt water's osmolarity was 5 OsM. Tomato plant cells have an omsolarity ~ 0.2 - 0.3 OsM. Clearly, the salt water was **hyper**osmotic to the plant cells. The salt water was **hyper**tonic also, as the microscopic video demonstrated (recall that the plant cells shrunk dramatically). Most plasma membranes have few channel proteins for Na^+ and Cl^-, so these cells have a very low permeability to these ions. Low permeability to salt (NaCl) would prevent the Na^+ and Cl^- solutes from diffusing into the cell. Instead, water moves out of the cell by osmosis. By definition, any solution that dehydrates a cell is hypertonic. Your turn to try!

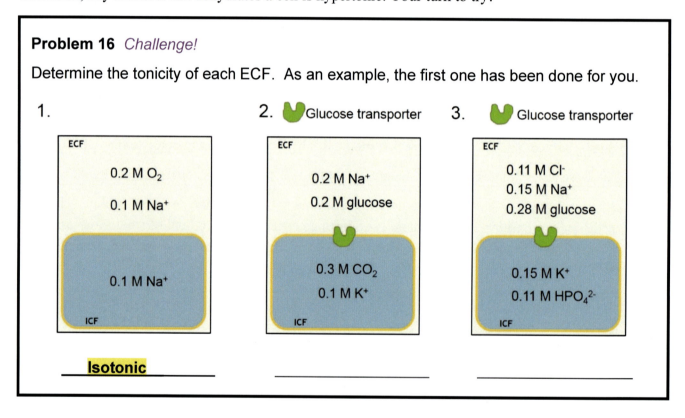

Problem 16 *Challenge!*
Determine the tonicity of each ECF. As an example, the first one has been done for you.

Tonicity is an important descriptor for solutions exposed to living cells. For example, an **intravenous** (**IV**) solution infused into a patient's blood must be **isotonic** to blood cells to prevent any catastrophic volume changes to the patient's blood cells. A 0.9% NaCl solution is both isoosmotic and isotonic to blood cells, and so 0.9% saline solution is a popular IV fluid (**Figure 10**). Another type of IV fluid that is used when the patient is low in blood sugar is D5 - 0.9% NaCl. This solution is 5% glucose + 0.9% NaCl.

---

[6] If you are still struggling with the concept of tonicity versus osmolarity, access the QR code found on the top of page 160 for another explanation.

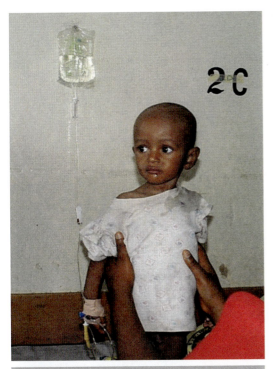

Figure 10. A Nigerian child receives an IV of isotonic fluid, possibly to rehydrate after life-threatening diarrhea.

If 0.9% saline is isoosmotic to blood cells, then 0.9% saline + extra glucose (5%) must be hyperosmotic to blood cells. Yet, the 5% glucose + 0.9% NaCl solution remains isotonic over the long haul. Why? Because most cells have glucose transporters embedded in their membranes, allowing glucose to diffuse into the cells. Thus, glucose does not stay concentrated in the blood. Consequently, water does not leave the cells.

We are almost done! Thus far, we've explored the passive processes of diffusion and osmosis. Both processes alleviate a concentration gradient and therefore are energetically favorable. *But what about creating a concentration gradient?*

Transport of a single molecule or ion *against* its concentration gradient (i.e. towards an area that is more concentrated) <u>requires</u> both energy "payment" and a carrier protein (a channel protein won't work). Carrier proteins that <u>create</u> a concentration gradient are often called pumps[7].

Creating a concentration gradient is a type of **active transport**. All molecular transport can be classified as either *active* or *passive* transport. Let's define those terms now.

Passive transport across the plasma membrane requires no energy "payment" by the cell and includes:

(1) *simple diffusion* – diffusion through the transient gaps in the lipid bilayer of a membrane,

(2) *facilitated diffusion* – diffusion facilitated by a channel or carrier membrane transporter, and

(3) *osmosis* – water diffusion either through aquaporins or gaps in the lipid bilayer of a membrane.

In contrast, active transport requires energy payment by the cell. ATP catabolism is one type of energy payment. When ATP pays directly for pumping, we call that **primary active transport**.

In primary active transport, ATP attaches to a carrier protein and phosphorylates (i.e. attaches its phosphate group) the membrane transporter. This causes a conformational change in the carrier protein that strongly attracts the solute to be pumped. When the phosphate group pops off, the carrier protein is induced to change shape again – this time releasing the molecule (or ion) into the more concentrated solution. The most important example of primary active transport is the **sodium-potassium pump**[8] (**Figure 11**).

The sodium-potassium pump spends 1 ATP molecule to first pump 3 Na^+ ions out of the cell, followed by pumping 2 K^+ ions into the cell. Watch the following tutorial for an animated explanation.

---

[7] The term "pump" should make sense - like a tire pump that requires energy to *pump* air into a more concentrated compartment (the tire).

[8] Any transporter that is involved with ATP is called an ATPase. Thus, you will often see the sodium-potassium pump called a sodium-potasium ATPase because the pump is powered by ATP catabolism.

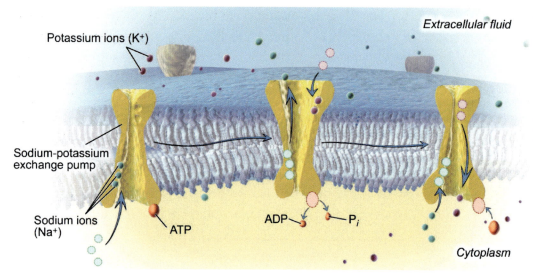

Figure 11. The Na^+-K^+ pump costs 1 ATP molecule and pumps both Na^+ and K^+ against their concentration gradients.

Image by Blausen.com staff. "Blausen gallery 2014". Wikiversity Journal of Medicine. Licensed under CC BY 3.0 via Wikimedia Commons.

Problem 17

Mark each statement about the sodium-potassium pump true or false. If the statement is false, explain why it is false.

1. Each cycle costs 2 ATP.

2. One cycle pumps three Na^+ ions to the ECF and two K^+ ions into the ICF.

3. The sodium-potassium pump is a channel protein.

4. The sodium-potassium pump is an example of primary active transport.

5. The sodium-potassium pump creates and maintains the concentration gradients of sodium and potassium across the plasma membrane.

6. The ions moved by the sodium-potassium pump are examples of facilitated diffusion.

The strong sodium and potassium gradients across human plasma membranes are maintained by the continual work of sodium-potassium pumps. These gradients are absolutely essential for the electrical conductions of the nervous system and muscles (including your cardiac muscle) to function properly in your body. If you want to know just how important those gradients are – chew on this morbid fact. The third injection in *death by lethal injection(s)* is straight potassium to the heart. The K^+ injection floods the ECF around the heart with potassium ions, which ruins the concentration gradient of K^+ established by the pump. The K^+ gradient is lost, causing cardiac arrest (i.e. the heart stops beating).

In addition to the pivotal role the sodium-potassium pump plays in muscle and nervous system physiology, the Na^+ concentration gradient is a form of potential energy that drives other types of active transport. Let me explain.

The sodium-potassium pump concentrates sodium in the ECF. There are several types of carrier proteins that allow Na$^+$ to diffuse back *into* the cell, but ONLY if another solute comes with it. A classic example of this is the sodium-glucose cotransporter. In **Figure 12**, glucose (the blue hexagon) is already concentrated in the cell. Moving even *more* glucose into the cell (against its concentration gradient) is active transport and requires energy payment. The Na$^+$- glucose cotransporter is designed to couple the energy *released* when sodium diffuses into the ICF to the energy *payment* required to transport glucose against its concentration gradient. In other words, sodium diffusion "pays" for pumping glucose into the cell. This type of glucose transport is **secondary active transport**, where the diffusion of one concentrated substance is coupled to the active transport of another substance. This is different than primary active transport, where ATP catabolism pays for active transport directly.

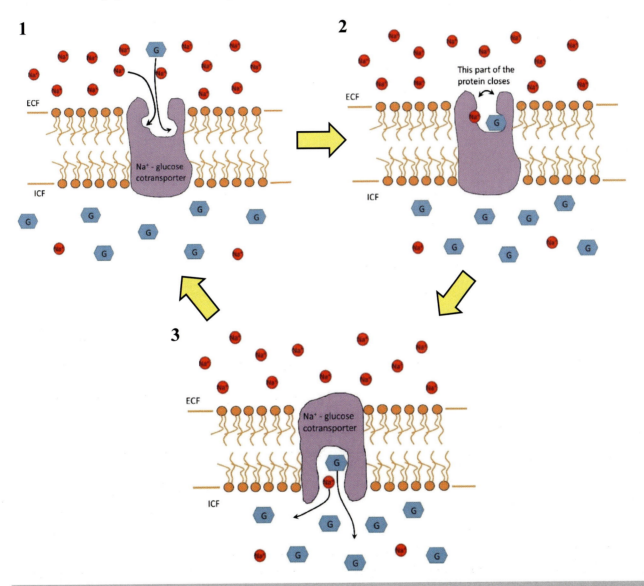

Figure 12. The sodium-glucose cotransporter allows the diffusion of sodium to drive the transport of glucose against its concentration gradient. (1) The cotransporter first binds a Na$^+$ ion (red circle), which is eager to get out of the ECF. (2) But in order for the carrier protein to change shape, a glucose (blue hexagon) must also bind. The binding of both Na$^+$ and glucose induces the conformational change in the cotransporter. (3) The cotransporter releases both Na$^+$ and glucose into the ICF.

It might seem like an unnecessary extra step to use the sodium gradient to push glucose against its concentration gradient. *Why doesn't the cell just use primary active transport? Like a carrier protein that spends 1 ATP to pump 1 glucose into the cell?* That's an excellent question. The answer lies in cost efficiency.

Using primary active transport, like a carrier protein that would pump 1 glucose across for 1 ATP payment[9], is too costly. Remember that the Na^+-K^+ ATPase pumps three Na^+ ions out at the cost of 1 ATP. Those three Na^+ ions can then diffuse back into the cell; each ion brings one glucose with it. So, by using the sodium-potassium pump first, and then a sodium-glucose cotransporter, you actually get to transport 3 glucose molecules into the cell for only 1 ATP. This is a more cost-efficient approach.

Good job on finishing this very important chapter. One more problem and you're done. ☺

Problem 18 *Challenge!*

The list below contains words or phrases that were introduced in this chapter. Classify each word or phrase as applicable to passive or active transport. Three of the terms are applicable to BOTH active and passive transport. As an example, one has been done for you.

PASSIVE TRANSPORT		ACTIVE TRANSPORT
Aquaporin		

Simple diffusion Na^+-K^+ *pump*   *Channel protein*

Primary active transport   *Endocytosis*   *Osmosis*

Facilitated diffusion   *Carrier protein* Na^+-*glucose cotransporter*

~~*Aquaporin*~~   *Secondary active transport*   *Exocytosis*

---

[9] Inquisitive students sometimes ask why a glucose pump would only be able to transport one glucose per cycle of the pump (i.e. one glucose is moved per ATP). After all, the sodium-potassium pump moves three sodium ions per cycle. The most probable answer is size. Na^+ is much smaller than glucose, so a carrier transporter can accommodate three small ions. Glucose is much bigger, and so the maximum number of glucose molecules that can fit into a carrier protein during one cycle is probably only one.

Chapter 11: Answer key

1. **d** is the *best* answer.   *Explanation*: a & c still show concentrated areas; b is perfectly uniform, which is unlikely to be realistic (molecules and ions don't usually retain a perfectly spaced out distribution as they move, even when diffusion equilibrium has been achieved).

2. 98°F (*higher temperature*); H^+ (*lower mass*); 50x more concentrated (*greater concentration gradient*).

3. O_2, CO_2, fatty acid (uncharged). The following QR code links to a detailed explanation.

4. O_2, CO_2, fatty acid (uncharged), steroid, H_2O most of the time

5. channel

6. 1. b   2. a

7. Any molecules larger than a monomer (simple sugar, amino acid, nucleotide), are probably too big to use protein-mediated transport and therefore must use vesicles for transport. If a large quantity of materials needs to be exported out or imported into the cell at the same time, a vesicle would be more efficient than carrier proteins. Also, any time the Golgi body is involved in packaging and shipping, the cell is likely to use exocytosis for secretion simple because vesicles bud from the Golgi body carrying the stuff.

8. ECF – <u>E</u>xtra<u>c</u>ellular <u>f</u>luid   ICF – <u>I</u>ntra<u>c</u>ellular <u>f</u>luid

9. To reduce the permeability to water, (1) more cholesterol should be added into the lipid bilayer and, (2) the aquaporins need to be removed.

10. 0.1M glucose + 0.1 M urea solution;   10 g/L glucose + 10 g/L maltose + 15 g/L KCl

11. 5% urea;   0.15 M glucose;   20 g/L glucose

12.

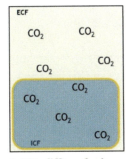 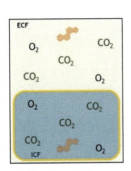 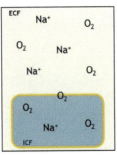 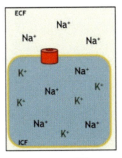

1. CO_2 diffuses freely through the phospholipid bilayer. The cell does not gain or lose water as the concentration of solute equalizes between the ICF and ECF.

2. All solutes diffuse freely through the phospholipid bilayer. The cell does not gain or lose water.

3. O_2 diffuses freely through the phospholipid bilayer but Na^+ does not. With more solute remaining in the ECF, water moves from the ICF to the ECF and the cell shrinks in volume.

4. K^+ cannot diffuse through the membrane but Na^+ diffuses through the sodium channel. The result – Na^+ reaches diffusion equilibrium but K^+ stays concentrated in the ICF. Water moves into the more concentrated ICF and the cell expands until the total solute concentrations are equalized in the ICF and ECF.

13.   12-2 is hypoosmotic; 12-4 is slightly hyperosmotic. A detailed explanation for problem 13 is available through the following QR code.

14. *Explanation:* Pure water, which is hypoosmotic (and hypotonic) to all cells, creates an osmotic gradient that draws water into the cell, expanding the cell volume. Cells with cell walls (e.g. plants, fungi and bacteria) are protected as the osmotic force (pressure from incoming water) is counteracted by the rigid force of the cell wall. This assumes the cell wall is structurally stable. The plant cell is slightly swollen as the plasma membrane pushes firmly against the cell wall, but the cell remains intact. Human cells lack a cell wall. The incoming water stretches the comparatively flimsy plasma membrane to the point of lysis (bursting). The result is cell death, where the hemoglobin parts are spilling out of the ruptured cell.

    Plant cell   Red blood cell

15. c) hypoosmotic

16. 1 - isotonic 2 - hypertonic 3 - isotonic. If your answers were not correct, a detailed explanation for how to correctly solve the problems can be accessed with the following QR code.

17. 2, 4 and 5 are true statements. Corrections for statements 1, 3 and 6 are as follows:

    1) Each cycle costs 2̶ 1 ATP.   3) The sodium-potassium pump is a c̶h̶a̶n̶n̶e̶l̶ carrier protein.   6) The ions moved by the sodium-potassium pump a̶r̶e̶ ̶e̶x̶a̶m̶p̶l̶e̶s̶ ̶o̶f̶ ̶f̶a̶c̶i̶l̶i̶t̶a̶t̶e̶d̶ ̶d̶i̶f̶f̶u̶s̶i̶o̶n̶ go against their concentration gradient so this is not diffusion; this is active transport.

18. Passive transport only – simple diffusion, facilitated diffusion, osmosis, aquaporins, channel proteins

    Active transport only – primary active transport, Na^+ - K^+ pump, endocytosis, exocytosis[10]

    Both passive and active – carrier proteins (some carrier proteins are used to allow diffusion and others create concentration gradients); Na^+ - glucose cotransporter and secondary active transport (in each case, one solute is *diffusing* down its concentration gradient and the other is moving against its concentration gradient).

---

[10] Endocytosis and exocytosis are classified as active transport because ATP energy is spent to move vesicles.

Chapter 12: Elegant enzymes

Understand enzymes and "-ase" (ace) your test.

A chemical reaction can occur when the potential energy stored in the reactants is greater than the products. Such reactions are **exergonic** or **spontaneous** reactions[1]. The term *spontaneous* should not be confused with *instantaneous*. A spontaneous reaction will happen at some point, but not necessarily right now. Case in point – consider this chemical reaction that summarizes aerobic respiration.

$$C_6H_{12}O_6 + 6\ O_2 \rightarrow 6\ CO_2 + 6\ H_2O$$

Glucose crystals in air

This reaction is highly exergonic and therefore spontaneous. But if you leave a jar of glucose exposed to O_2, it doesn't react immediately to form carbon dioxide and water. *Why not?* Because of a little something called **activation energy**. In a chemical reaction, (1) existing covalent bonds are broken and (2) new covalent bonds are formed. It takes a little spark of energy to pop electrons out of their existing covalent bonds and into one or more new covalent bond(s). This "spark" of energy needed to get the chemical reaction rolling is called the activation energy.

In order to break existing covalent bonds, reactants have to interact in a specific manner. Essentially, the reactants have to collide with <u>enough speed</u> and at the <u>correct orientation</u> to facilitate the transfer of electrons from one pair of atoms to another. The following video illustrates this principle.

As you may infer, the activation energy is a barrier that prevents many chemical reactions from proceeding; this is the case with glucose failing to react instantly with O_2 in the air.

Problem 1

What is the best definition of **activation energy**?

a) The amount of energy required to break existing covalent bonds and promote the formation of new covalent bonds.

b) The amount of heat required to turn an exergonic reaction into an endergonic reaction.

c) Whether or not kinetic energy is required for a chemical reaction to proceed.

Problem 2

True or *false*? All chemical reactions require activation energy.

---

[1] You may want to review Chapter 10 to refresh on this terminology.

How does a chemical reaction overcome this activation energy barrier? In the previous video, you learned that the kinetic energy of motion is one option, but overcoming the activation energy barrier this way requires that (1) enough heat is applied to the system to increase the likelihood of collisions, and (2) the reactants are oriented correctly. In living organisms chemical reactions need to happen *on demand*, so waiting for the ~~planets~~ molecules to align by chance is just not practical. Instead, cells use large proteins called **enzymes** to lower the activation energy of chemical reactions, thereby making *spontaneous* chemical reactions practically *instantaneous*. The following analogy explains how enzymes "cheat" the activation energy issue.

Figure 1. My son climbs up a skeeball ramp hoping to "cheat" his way to a higher score by pushing the skeeballs directly in the holes.

Skeeball is an arcade game where you roll wooden or plastic balls up a ramp. The goal is to jump the ball off the ramp and into one of the holes (**Figure 1**). The smaller the hole, the more points you score. The ball must be released from your hand at a *very precise* speed and angle to make it into the smallest, high value hole (white arrow in **Figure 1**). Release the ball with the wrong speed or at an imperfect angle, and the ball will simply bounce out of that tiny hole.

Releasing the skeeball from your hand at just the right speed and angle to make it into a high-value hole is analogous to how the reactants must collide to overcome the activation energy barrier. That is, they must collide with precise speed and orientation. If you're an average skeeball player (like me), skeeballs rarely fall into that high-value hole. *But what if you were to cheat?* Imagine climbing up the ramp with the balls tucked in your sweater until you are less than an arm's reach away from that small hole. Why, you could drop the skeeballs directly into the small hole and never miss! That is exactly what my 2-year-old son is attempting to do in **Figure 1**[2].

Enzymes are large proteins that "cheat" the activation energy issue similar to how climbing up the ramp cheats the game of skeeball. The following video explains *how* enzymes lower the activation energy.

In summary, enzymes are large proteins that lower the activation energy of a chemical reaction by

(1) attracting reactants to the enzyme,

(2) stressing existing covalent bonds so they break, and

(3) correctly orienting the reactants so that new covalent bonds are likely to form.

---

[2] Many people cheated this way to get more tickets from an arcade. In **Figure 1**, the arcade has wisely installed plexiglass over the holes to prevent kids (or adults) from cheating in this manner.

Simply, enzymes bring the reactants together, exert some "muscle" to break the existing covalent bonds and keep the reactants together in an optimum configuration so new covalent bonds form. From a functional standpoint, enzymes take spontaneous chemical reactions and make them practically instantaneous. In fact, enzymes are so good at what they do, a single enzyme can **catalyze** (i.e. perform) over a thousand chemical reactions per second!

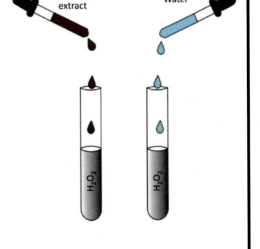

Problem 3

As described in the previous video, the enzyme **catalase** catalyzes this chemical reaction:

$$2 H_2O_2 \rightarrow 2 H_2O + O_2$$

I have two test tubes containing a 3% hydrogen peroxide solution. To the left test tube, I add liver extract which contains catalase. To the right test tube, I add water.

1. Which tube will have the least amount of H_2O_2 after thirty seconds?
   a) left tube   b) right tube   c) both tubes will still have about 3% H_2O_2

2. Which tube will have produced the most O_2 after thirty seconds?
   a) left tube   b) right tube   c) both tubes will contain equal amounts of O_2

Functioning enzymes are a requirement to stay alive. Without enzymes, cells cannot run chemical reactions reliably and therefore die. Since enzymes are so crucial to a cell's livelihood, the remainder of this chapter explores enzyme structure and function.

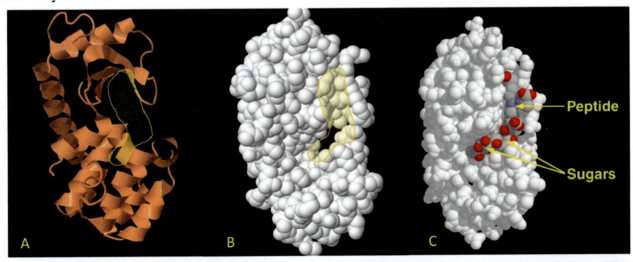

Figure 2. Three views of the same enzyme (lysozyme). (A) Folded ribbon model showing alpha helices with active site outlined in light yellow. (B) Lysozyme model from same angle. (C) Lysozyme with substrate shown embedded in the active site. Protein Images A and B created from PDB entry 1LYD in www.rcsb.org

Enzymes are large proteins. Each enzyme has a groove where the chemical reaction takes place; this groove is called the **active site** (**Figure 2**). The reactants that are chemically altered by the enzyme also get a new name – **substrates**.

Problem 4

This image depicts the enzyme pepsin, which cuts peptides up into individual amino acids in your stomach.

1. Circle the active site on this enzyme.

2. Name the *substrate* for this enzyme using the information provided in this problem.

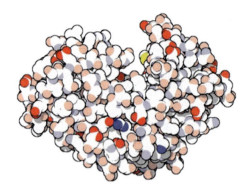

Image attributed to David S. Goodsell and RCSB PDB.

Since enzymes are large proteins, you should recall from Chapter 7 that the three dimensional structure of an enzyme is critical to its function. Each enzyme is optimally shaped to run a specific chemical reaction. In particular, the exact shape and chemical affinity of the active site is crucial. The active site must attract, fit and orient the substrates precisely. Consequently, enzymes are highly specific, much like a lock is highly specific for one key. Usually, each enzyme performs one and only one type of chemical reaction. If the cell has 200 different chemical reactions that it wants to perform, then we expect the cell to make 200 different types of enzymes. To give you a better handle on this "one enzyme = one chemical reaction" guideline, look carefully at **Figure 3**, which shows four *different* enzymes all working on the same substrate.

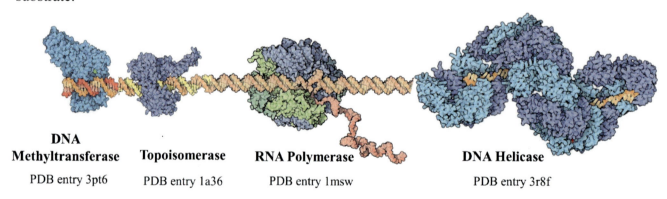

DNA Methyltransferase	Topoisomerase	RNA Polymerase	DNA Helicase
PDB entry 3pt6	PDB entry 1a36	PDB entry 1msw	PDB entry 3r8f

All images attributed to David S. Goodsell and RCSB PDB.

Figure 3. Four different enzymes (shown in blue or green) are needed to do four different things with DNA (the orange double helix). For example, DNA methyltransferase adds a methyl ($-CH_3$) group to DNA, while topoisomerase allows some of the knots that develop when DNA hypercoils to unwind. Note that each enzyme has a unique and specific 3D shape, reflecting that each enzyme performs a different chemical reaction with DNA.

Problem 5

Look carefully at the names of the enzymes in **Figure 3**. What two general conclusions can you make about enzymes? Circle both correct answers.

a) Enzymes often have names that end in - *ase*.
b) One substrate can only be acted on by a single enzyme.
c) One enzyme can run many different types of chemical reactions.
d) The name of the enzyme often describes the chemical reaction that enzyme is designed to catalyze.

Whenever an enzyme catalyzes (performs) a chemical reaction, it is customary to write the name of the enzyme over the arrow in the chemical reaction. For example, **hexokinase** is the enzyme that takes glucose and changes it into glucose 6-phosphate[3] as shown in **Figure 4**. Notice how much larger the enzyme is compared to its substrates.

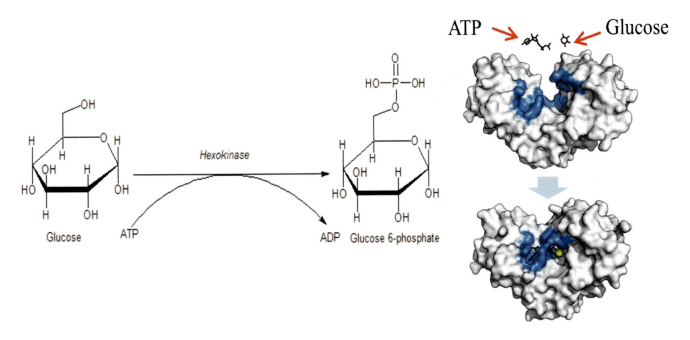

Figure 4. The enzyme hexokinase catalyzes this reaction: Glucose + ATP → Glucose 6-Phosphate + ADP. The left image shows the chemical reaction with the name of the enzyme written over the arrow. The right image shows the 3D structure of hexokinase (white and blue) before binding the substrates (top) and during the actual chemical reaction (bottom) with the substrates nestled into the active site (blue area). Right image is "Hexokinase induced fit 2" by Thomas Shafee. Licensed via CC at Wikimedia Commons.

Use the information in **Figure 4** to answer Problem 6.

---

[3] Glucose ($C_6H_{12}O_6$) has six carbons. In higher level biology courses, it is customary to number the carbons to keep track of what is happening to each one. The name glucose 6-phosphate refers to the phosphate group attached to the "6th" carbon.

Problem 6

1. Write the chemical reaction catalyzed by hexokinase.

2. Name the <u>two</u> substrates for the enzyme hexokinase.

3. One product produced by the enzyme hexokinase is glucose 6-phosphate. What is the *other* product?

4. In **Figure 4**, circle the active site of hexokinase.

5. In **Figure 4**, put a box around the phosphate group in the chemical structure of glucose 6-phosphate.

A single enzyme will perform the same chemical reaction over and over again, as long as (1) the active site is *"open for business"* and (2) the substrates are readily available. Let's use this information to explore the reaction rate. We will use the enzyme catalase which performs the following chemical reaction – $2\ H_2O_2 \rightarrow 2\ H_2O + O_2$ – as our model.

"Colourful wrapped salt water taffy" by Lara604. Licensed under CC via Wikimedia Commons.

The **reaction rate** is defined by how quickly a particular chemical reaction proceeds. In practical terms, this is measured by <u>how much product is formed per unit of time.</u> As an analogy, imagine if you had a job wrapping saltwater taffy pieces with wax paper. You would be the enzyme (let's call you *taffywrapperase*). Your "substrates" would be a piece of taffy and a square of wax paper. The "product" of your chemical reaction would be the wrapped taffy. The <u>reaction rate</u> would be defined by how many pieces of wrapped taffy you produce per minute.

How many pieces of wrapped taffy *can* you produce per minute? That answer depends on a number of factors. For instance, imagine if the taffy pieces were coming off a conveyer belt at a slow rate – only one piece of taffy per minute. Your reaction rate would be one wrapped piece of taffy per minute. What if the taffy pieces were released every 30 seconds onto a conveyer belt? Then your reaction rate is two pieces of wrapped taffy per minute. What if the taffy pieces were released every 10 seconds? Your reaction rate could increase to six pieces of wrapped taffy per minute. What if the taffy pieces were released every second? Wait. Are you physically able to wrap 60 pieces of taffy in a single minute? Probably not.

This example illustrates that the reaction rate can depend upon the frequency that the enzyme (you) encounters its substrate(s) (the taffy and wax paper). At some point, however, the active site (your hands) is working as fast as it can. This is the point at which the enzyme is **saturated**. To be *saturated* means to be *filled to capacity*. Using the taffy analogy, I expect that most people are saturated somewhere around 10 – 15 pieces of taffy per minute. Even if 20 pieces of taffy were released per minute onto a conveyer belt, a single taffywrapperase enzyme (i.e. person) is never going to produce 20 pieces of wrapped taffy in a minute (**Figure 5**).

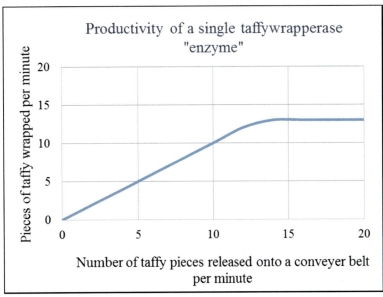

Figure 5. A line graph showing the reaction rate of a single hypothetical *taffywrapperase* enzyme. The more frequently the enzyme encounters its substrate (taffy pieces), the more product is produced until the enzyme is saturated. Real enzymes function the same way; the reaction rate is dependent upon the frequency that the enzymes encounter the substrates.

Problem 7

Interpreting graphs is a necessary skill that allows you to comprehend data quickly. Use the graph in **Figure 5** to answer the following questions. The first one has been done for you as an example.

1. If 5 taffy pieces are dropped onto a conveyer belt per minute, what is the reaction rate?
   _____5_____ pieces of wrapped taffy/minute.

2. If 10 taffy pieces are dropped onto a conveyer belt per minute, what is the reaction rate?
   _____ pieces of wrapped taffy/minute.

3. If 20 taffy pieces are dropped onto a conveyer belt per minute, what is the reaction rate?
   _____ pieces of wrapped taffy/minute.

4. In **Figure 5**, identify the point at which *taffywrapperase* is saturated.

All enzymes are subject to saturation. Assume a single enzyme is saturated when producing 500 products per second. What if the cell needs 2000 products per second? Obviously one enzyme isn't enough. The solution is for the cell to add more enzymes! Adding more enzymes is like adding more people to wrap taffy coming off the conveyer belt. The more people wrapping taffy, the more wrapped taffy can be produced per minute. Likewise, adding more enzymes will increase the reaction rate of a chemical reaction.

What *else* can affect the reaction rate? We can summarize the factors that affect reaction rate into three groups. The reaction rate is governed by:

(1) the number of available enzymes,

(2) the frequency an enzyme encounters its substrate(s), and

(3) the active site's capacity to perform the chemical reaction.

I have already touched on the first parameter; more enzymes equals more products produced per unit of time. We'll use the remainder of the chapter to explore the frequency that an enzyme encounters its substrate(s), and the active site's capacity to perform the chemical reaction.

Going back to the taffy-wrapping analogy – when the taffy was released from a conveyer belt at one piece/minute, the wrapping rate was capped at 1 wrapped taffy/minute. Similarly, if an enzyme encounters its substrate once per minute, then the reaction rate will be 1 product/minute. **The more frequently an enzyme encounters its substrate, the higher the reaction rate.** *What determines how often an enzyme finds its substrate(s)?* Two factors influence how frequently a single enzyme encounters its substrate(s): temperature and concentration of the substrate.

The higher the temperature, the faster molecules move and the less time it takes before an enzyme encounters its substrate. Faster molecular movement therefore increases the frequency that an enzyme's active site meets its substrates. In turn, this increases the chemical reaction rate. The converse is also true. In colder environments, molecules move more slowly and thus reaction rates are also lower. This is why refrigerators are useful in prolonging the shelf-life of food contaminated with bacteria.

The concentration of substrate also matters. If there are lots of substrate molecules, then the enzyme's active site is likely to find substrates quickly and the reaction rate is high. Conversely, if the substrate concentration is low, then it may take a while before the enzyme encounters a single substrate molecule. The result is a lower reaction rate. The following video summarizes these points.

Problem 8

Catalase catalyzes this chemical reaction: $2\ H_2O_2 \rightarrow 2\ H_2O + O_2$.

1. List the products of this chemical reaction. _____, _____

2. A student mixes hydrogen peroxide and catalase together in a test tube. To increase the amount of O_2 produced per second, the student could… (circle all correct answers)

   a) run the chemical reaction at 4°C rather than at room temperature.
   b) increase the concentration of H_2O_2 in her test tube.
   c) add more catalase enzymes to her test tube.

Problem 9

A student runs the chemical reaction described in Problem 8 at two different temperatures – 10°C and 25°C. The student displays her data with the following graph.

Whoops! The student forgot to label the lines.

Which line represents the reaction at 10°C? Which line represents the reaction at 25°C?

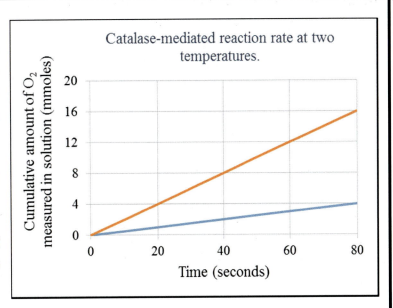

Catalase-mediated reaction rate at two temperatures.

The final parameter that governs the rate of chemical reactions is whether or not the active site is "open for business". To perform a chemical reaction, the enzyme's active site must be structured to attract, hold and chemically alter the substrates. That's not always the case.

Here are some things that could shut down the active site to stop the enzyme from performing a chemical reaction.

(1) Another molecule could get lodged into the active site, blocking the substrate from binding with the active site.

(2) The active site could change shape so the substrates no longer fit into the enzyme's active site.

(3) The active site might be missing critical pieces.

On a side note......

Spoilage is the result of contaminating bacteria or fungi growing in the food. At 4°C, refrigerating meats and other moist foods allow foods to spoil more slowly. A cold refrigerator slows down molecular motion, which means bacterial enzymes encounter their substrates less frequently. The slower reaction rates hinders bacterial growth and reproduction. Thus, slowing down the rate of chemical reactions slows down the rate of bacterial growth. Food stays edible longer.

It rarely makes sense for all enzymes to be running 24/7. For example, if DNA polymerase – the enzyme that replicates your DNA – had an active site that was always *open for business*, your chromosomes would be in a constant state of replication even if the cell had no intention of ever dividing. What a waste of energy that would be! Thus, many enzymes are designed to be regulated. By regulated, I mean that enzymes can be turned *off* and *on* as a cell needs. An enzyme that is *on* would have a properly structured active site – one that attracts, binds and acts on its substrates with ease. An enzyme that is *off* has a closed or unavailable active site (**Figure 6**).

Figure 6. The purple enzyme is **on** when the active site can bind the substrate (orange) and produce the products (red). The enzyme is **off** when the active site cannot bind the substrate and produce the products. Top right – another molecule is lodged in the bottom of the enzyme which closes the active site. Middle right – something else (green molecule) is in the active site preventing access by the substrate. Lower right – a piece is missing from the active site.

Enzymes that toggle between on/off states are advantageous for organisms.

For example, cells of your stomach lining secrete the enzyme pepsin into the stomach cavity. Pepsin is an enzyme that digests proteins, cutting polypeptides and proteins into shorter peptides and amino acids. When synthesizing pepsin, how can stomach cells prevent pepsin from digesting its own precious proteins? The answer is this: the cell keeps the pepsin enzyme in the *off* position until the enzyme is secreted out of the cell (**Figure 7**).

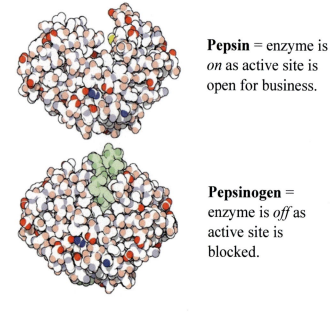

Pepsin = enzyme is *on* as active site is open for business.

Pepsinogen = enzyme is *off* as active site is blocked.

The fit between the active site and substrates can be conceptualized as two puzzle pieces fitting together. What happens if the enzyme encounters a molecule that is structurally similar to its substrate(s)? It's possible that such a molecule would mimic the substrate and bind to the active site as shown in **Figure 6** (middle right image). Any molecule that is not the substrate but can still wedge itself into the active site for any length of time is called a **competitive inhibitor** because it *competes* with the true substrate for access to the active site. Many therapeutic drugs are competitive inhibitors that shut down or reduce the activity of a particular enzyme.

Image attributed to David S. Goodsell and RCSB PDB.

Figure 7. Pepsin is the active form of the enzyme; pepsinogen is the inactive form of the enzyme. Stomach lining cells synthesize and secrete *pepsinogen*. This strategy is intentional; any active pepsin enzyme would cut up the cell's own proteins and destroy the cell. Once pepsinogen is secreted, the amino acids in green are removed (by a process not described here), exposing the active site of pepsin. Now, the enzyme is *on* and ready to start cutting up the proteins you swallowed in your food.

Problem 10

Bacteria secrete the enzyme *transpeptidase* which synthesizes peptidoglycan – a large structure that strengthens bacterial cell walls. Penicillin is a competitive inhibitor for transpeptidase. When bacteria are exposed to penicillin, what is the outcome?

a) The bacterial cell walls will be weaker.
b) The bacterial cell walls will be stronger.
c) The bacterial cell walls will remain the same as before.

Enzymes that are turned on and off frequently often have an **allosteric site**. The allosteric site, which is distinct from the active site, is another groove that attracts a *regulatory* molecule. The regulatory molecule turns the enzyme on or off by "remotely" controlling the shape of the active site (**Figure 8**).

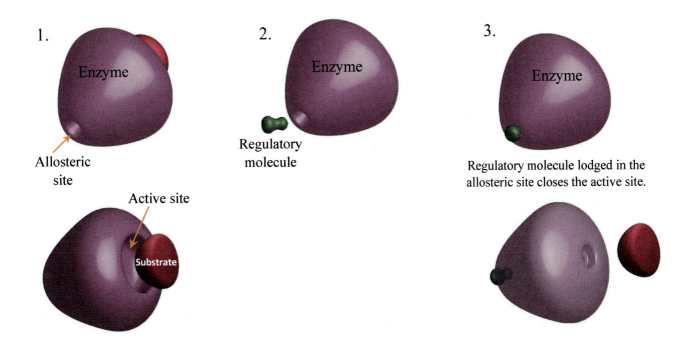

Figure 8. Allosteric sites allow for regulation of an enzyme by a chemical other than the substrate – a regulatory molecule. (1) The allosteric site is a groove distinct from the active site. In this enzyme, the active site fits the substrate when the allosteric site is empty. (2) A regulatory molecule approaches the allosteric site of the enzyme. (3) When the regulatory molecule binds to the allosteric site of the enzyme, the active site of the enzyme changes shape. The substrate no longer fits into the active site, so the enzyme cannot perform the chemical reaction.

In **Figure 8**, the regulatory molecule is called an **allosteric inhibitor** because the enzyme is inhibited when the allosteric site is occupied by the regulatory molecule. But not all regulatory molecules turn enzymes off. In fact, some enzymes with an allosteric site are actually turned *on* by their regulatory molecule. When the regulatory molecule binds to the allosteric site, the active site becomes *open for business* and the chemical reaction proceeds. If an enzyme operates that way, then the regulatory molecule is known as an **allosteric activator**. Note that the terminology tells you about the purpose of the regulatory molecule.

Allosteric inhibitors <u>inhibit</u> (turn *off*) an enzyme and allosteric activators <u>activate</u> (turn *on*) an enzyme when bound to the allosteric site.

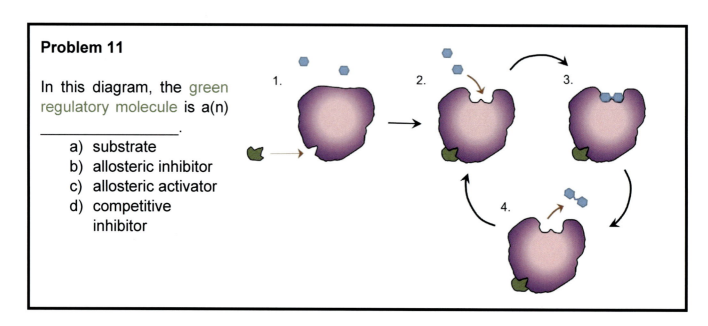

Problem 11

In this diagram, the green regulatory molecule is a(n) _____.

a) substrate
b) allosteric inhibitor
c) allosteric activator
d) competitive inhibitor

Let's recap. We've seen that other molecules can affect an enzyme's active site by binding to either the active site or the allosteric site. *Are there other strategies a cell can employ to govern enzyme activity?* You bet! Enzymes are large proteins - long polypeptides folded into a specific 3D shape. While a protein is always the major component in every enzyme, it is not always the only component. Many, enzymes require an additional piece inserted before the enzyme is functional (**Figure 9**). The additional piece is called a **cofactor**. Cofactors can be inorganic (such as a metal ion) or organic. If the cofactor is organic, it may go by another name – **coenzyme**. Coenzymes are often a modified vitamin[4]. NAD and FAD – your friendly taxicabs from Chapter 10 – are actually coenzymes.

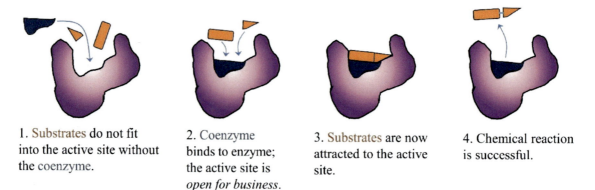

1. Substrates do not fit into the active site without the coenzyme.
2. Coenzyme binds to enzyme; the active site is *open for business*.
3. Substrates are now attracted to the active site.
4. Chemical reaction is successful.

Figure 9. Some enzymes are designed to be inactive without a **cofactor** completing the active site.

---

[4] If you've ever wondered why vitamins are so important, you now have the answer. Many vitamins are precursors to coenzymes. Essentially, vitamins are needed to keep your enzymes running.

Problem 12

This bacterial enzyme requires two magnesium ions (Mg^{2+}) to function (shown in green circles). For this enzyme, magnesium is a _____.

  a) cofactor only
  b) coenzyme only
  c) cofactor and coenzyme

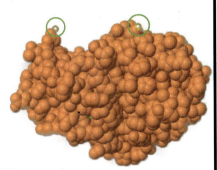

Protein image created from PDB entry 1HW6 in www.rcsb.org

Problem 13

Match each enzyme image to the correct description. As an example, one has been done for you.

1. Enzyme works normally: __B__
2. Enzyme inactivated by an allosteric inhibitor: _____
3. Enzyme missing a cofactor: _____
4. Enzyme shut down by a competitive inhibitor: _____

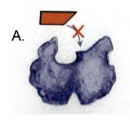
A.
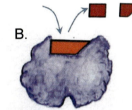
B.
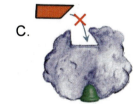
C.
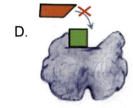
D.

Because enzymes usually participate in a collective effort known as a metabolic pathway, a cell may only need to turn off one enzyme in the pathway to shut down the entire pathway. For example, let three enzymes be involved in the conversation of substrate A to reactant D.

$$A \xrightarrow{\text{Enzyme 1}} B \xrightarrow{\text{Enzyme 2}} C \xrightarrow{\text{Enzyme 3}} D$$

If the cell's endgame is to make a bunch of D, then all three enzymes are required. *Enzyme 1* takes A and makes B. Then *enzyme 2* takes B and makes C. Finally, *enzyme 3* converts C into D. If a cell wants to stop making D, then it only needs to turn off **one** of the enzymes. In Human Physiology, you will encounter this scenario frequently. A metabolic pathway might involve dozens of enzymes, but only one enzyme is turned *on* and *off* routinely to regulate the pathway.

Some therapeutic drugs are enzyme inhibitors that shut down only one enzyme in a pathway. For example, Viagra® is a drug that maintains penile erection. It might surprise you to learn that Viagra® is actually an enzyme inhibitor; it blocks an enzyme in a pathway that functions to restrict blood flow into the penis. The result is continual blood flow into the penis resulting in sustained engorgement. Antabuse is another therapeutic drug that competitively inhibits one of two enzymes in the alcohol detoxification pathway. Read on to learn something useful. ☺

The alcohol detoxification pathway converts ethanol into harmless acetic acid (you know this as vinegar) as shown below. Let's say you consume copious amounts of ethanol[5]. As ethanol is absorbed by your intestinal tract, your blood alcohol level rises. This slows down reaction time, makes coordination difficult, and releases your inhibitions. You're drunk. *Alcohol dehydrogenase* enzymes in your liver encounter ethanol and begin changing it into acetaldehyde; acetaldehyde is poisonous and contributes to sickness and at elevated levels, induces vomiting. As your acetaldehyde levels rise a few hours after drinking, you'll experience the distinctly unpleasant sensation of a hangover. But it won't last forever because *aldehyde dehydrogenase* converts the toxic acetaldehyde into acetic acid, which is harmless. As acetaldehyde levels drop, the symptoms of the hangover subside.

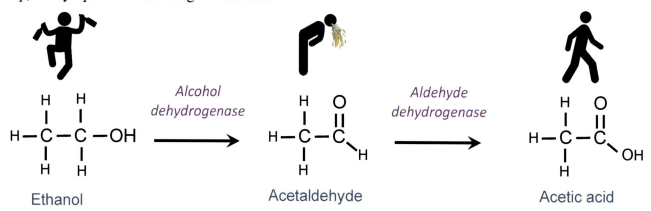

Problem 14

1. Antabuse is a drug that competitively inhibits aldehyde dehydrogenase. Thus, Antabuse is a(n) _____ for aldehyde dehydrogenase.

    a) substrate   b) allosteric inhibitor   c) allosteric activator   d) active site inhibitor

2. Two identical twins (let's call them Ann and Beth) both drink five shots of vodka. Ann takes an Antabuse pill with the five shots of vodka. Beth does not. After four hours, what statement is true?
    a) Ann's ethanol levels will be much higher than Beth's ethanol levels.
    b) Ann's acetaldehyde levels will be much higher than Beth's acetaldehyde levels.
    c) Ann's ethanol levels will be much lower than Beth's ethanol levels.
    d) Ann's acetaldehyde levels will be much lower than Beth's acetaldehyde levels.

In Chapter 10, we learned that only exergonic reactions are energetically favorable. Yet, many chemical reactions are fundamentally endergonic; the reactants have less chemical energy than the products[6]. Many students are unsure about the role of enzymes in endergonic reactions.

---

[5] This is strictly a hypothetical situation; not advice or a suggestion of any type.
[6] Review pages 122-123 specifically for a review of exergonic and endergonic reactions.

Are enzymes involved in these nonspontaneous, endergonic reactions? If so, do enzymes supply the energy "payment" needed? Let's address these excellent questions.

1. Do enzymes catalyze nonspontaneous (endergonic) chemical reactions? Yes, absolutely!

2. Do enzymes "pay" for the energy difference needed to run an endergonic chemical reaction?

No, they do not. Do not confuse *activation energy* with *chemical energy* constrained by the laws of thermodynamics. Enzymes – which do not actually participate in any chemical reaction – do NOT pay for the energy difference to run endergonic chemical reactions.

In Chapter 10, you learned that the highly exergonic ATP catabolism is often coupled with endergonic reactions as a method of "payment". If ATP is the payment, then think of enzymes like cashiers. Enzymes *accept* ATP payment to run an endergonic reaction. The enzyme doesn't pay for the energy deficit directly, but rather cashes in the energy released from a highly exergonic reaction (ATP → ADP + P) to pay for an endergonic one.

ATP's involvement with enzymes extends beyond powering endergonic reactions. ATP is often used as a regulatory molecule. For example, ATP is capable of phosphorylating some enzymes. Phosphorylation can alter the active site of the enzyme, thus controlling if the enzyme is *on* or *off*. **Figure 10** describes an unusual case of where a chemical reaction is technically spontaneous (energetically favorable), but the activation energy is so high that the enzyme by itself can't lower it to feasible levels. Instead, the enzyme spends 16 ATP to help lower the activation energy!

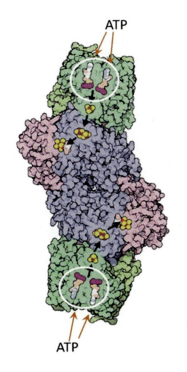

Image attributed to David S. Goodsell and RCSB PDB.

Figure 10. The enzyme nitrogenase is found in specialized bacteria that run the following chemical reaction: $N_2 + 8 H^+ + 8e^- \rightarrow 2 NH_3 + H_2$. N_2 is held together by a triple covalent bond that is very tight and extremely difficult to bend. Consequently, the activation energy needed to break the triple covalent bond is very great – greater than any enzyme's ability to muscle it.

In living systems, 16 ATP are cashed in by the enzyme nitrogenase to create highly reactive electrons capable of attacking the triple bond to reduce atmospheric nitrogen (N_2) to ammonia (NH_3). In this image, note the four ATP shown in the white circles. The reactive electrons are needed to weasel their way into the very strong triple covalent bond and weaken it.

Does this chemical reaction ever happen in nonliving systems? Yes, but only during lightning storms; the bolt of lightning carries enough energy to overcome the astronomical activation energy of this reaction.

Two more problems and you're done! Congratulations on making it to the end of the book. ☺

Problem 15

Which statement is false?

a) Enzymes can couple the catabolism of ATP with endergonic reactions.

b) Some enzymes are regulated by phosphorylation. ATP's third phosphate group attaches to the enzyme and changes the active site.

c) Enzymes can directly pay for the energy needed to fund endergonic reactions.

Problem 16 *Challenge!*

Look again at **Figure 4** and Problem 6. What role does ATP play with the enzyme hexokinase?

a) ATP is an allosteric activator.

b) ATP is an allosteric inhibitor.

c) ATP catabolism is paying for an endergonic chemical reaction involving glucose.

d) ATP is phosphorylating the enzyme and closing the active site.

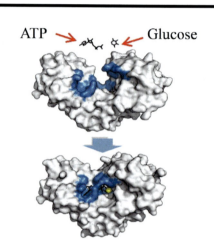

Chapter 12: Answer key

1. A

   Explanation: B is false because whether a reaction is endergonic or exergonic is strictly a property of the collective chemical energy stored in the molecules. C is false. Although motion is a form a kinetic energy, all chemical reactions have some sort of activation energy, so defining activation energy as requiring kinetic energy or not is false.

2. True

   Explanation: The quantity of energy required to activate a chemical reaction can vary, but all chemical reactions have at least a small amount of activation energy.

3. A is the correct answer for both questions. With catalase added to the left tube, the H_2O_2 will quickly be converted to O_2 and H_2O. This means that the amount of H_2O_2 will drop in the left tube as the concentration of O_2 increases.

4. The substrate is a peptide (water is also a substrate, for breaking a polymer into its monomers is an example of a hydrolysis reaction).

5. A, D

6. 1) Glucose + ATP $\xrightarrow{\text{Hexokinase}}$ Glucose 6-Phosphate + ADP
   2) ATP, Glucose
   3) ADP
   4 & 5) See images to the right

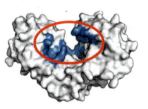
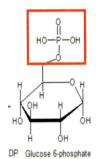

7. 2) 10;   3) 13;   4) Taffywrapperase is saturated at 13 pieces of taffy released onto a conveyer belt/minute.

8. 1. H_2O;   2. B, C are both correct answers

9. Orange line is 25°C; blue line is 10°C

10. A

11. C

    Explanation: If the green molecule was a substrate, it would fit into the active site (where the blue sugars go. Also, we would not use the phrase regulatory molecule to describe a substrate. If the green molecule were a competitive inhibitor, it would bind to the active site rather than the allosteric site. You can tell the difference because the active site is actually performing a chemical reaction; the allosteric site is not. If the green molecule were an allosteric inhibitor, we would expect the enzyme's active site to be closed up when the green molecule was bound in the allosteric site. Instead, we see that when the regulatory molecule binds to the allosteric site, the active site opens and begins performing the chemical reaction. This makes the green regulatory molecule an allosteric activator.

12. A

   Explanation: Mg^{2+} is inorganic. Since it is required to attract the substrates to the active site, it is a cofactor. If it were organic (i.e. with a carbon atom), it would be both a cofactor and a coenzyme.

13. 2 – C;   3 – A;   4 – D

14. 1. D   2. B

   Explanation: Any molecule that is a competitive inhibitor must bind to the active site and therefore is an active site inhibitor. If someone were to drink five shots of vodka plus an Antabuse pill, the aldehyde dehydrogenase enzyme would be blocked. Ethanol would still be converted to acetaldehyde normally (the alcohol dehydrogenase enzyme is unaffected). However, aldehyde dehydrogenase's active site is blocked, so the enzyme is unable to convert acetaldehyde into acetic acid. This means the acetaldehyde levels would be higher than a person consuming five shots of vodka without the Antabuse pill.

15. C

16. C

   Explanation: Although ATP is often involved in allosteric sites, you can see from these images that ATP is actually a substrate for the chemical reaction. Thus, ATP catabolism funds the conversion of glucose into glucose 6-phosphate. Glucose 6-phosphate has more chemical energy than glucose for reasons beyond the scope of this tutorial.

Web links for videos

The url link for the introduction video is: htttps://vimeo.com/133235810

Chapter 1 videos are hosted at this web address: https://vimeo.com/album/3528580

Chapter 2 videos are hosted at this web address: https://vimeo.com/album/3627315

Chapter 3 videos are hosted at this web address: https://vimeo.com/album/3627337

Chapter 4 videos are hosted at this web address: https://vimeo.com/album/3627346

Chapter 5 videos are hosted at this web address: https://vimeo.com/album/3627352

Chapter 6 videos are hosted at this web address: https://vimeo.com/album/3632827

Chapter 7 videos are hosted at this web address: https://vimeo.com/album/3632829

Chapter 8 videos are hosted at this web address: https://vimeo.com/album/3632831

Chapter 9 videos are hosted at this web address: https://vimeo.com/album/3632832

Chapter 10 videos are hosted at this web address: https://vimeo.com/album/3632833

Chapter 11 videos are hosted at this web address: https://vimeo.com/album/3632834

Chapter 12 videos are hosted at this web address: https://vimeo.com/album/3632835

Acknowledgements

I thank the following individuals for their contributions, assistance, and support with this book.

First and foremost, I thank my husband **Kacy Williams** and my two children who selflessly put up with an absentee mother and wife for almost a full year while I crafted this book.

Second, I thank **Nalini Asha Biggs** for spending many hours on illustrations, filming and editing, and providing guidance with the cover design. **Benjamin Brewer** was kind enough to be a guinea pig and reviewed almost all of the chapters with a nursing student's perspective.

In no particular order, I thank Leslie Thompson, Winston Williams, Mary Williams, Angelina Saunders, Carey Carpenter, Natasha Ramme, Chelsea Movilla-Diago, and Kacy Williams for offering editing suggestions and corrections.

The following individuals agreed to be subjects in videos or photos: Laketa Ducat, Gene Gushansky, Richard Albistegui-Dubois, Kenna Hanel, Rob and Leah Gonzalez, Benjamin Brewer, Beth Pearson and progeny, Mike and Valerie Deal and progeny, Cathy Jain and progeny, and Kacy Williams.

I thank Gordon Stevens for his videography of children "diffusing" and phospholipids. The Palomar TV Center kindly filmed several studio shots, including the introduction, the electron transport chain, and dissolving salt sequences. Finally, none of this would have been possible without the support of Palomar College who graciously provided me with a sabbatical to write the first half of this book.

Image credits

All photos, images, graphics and illustrations are the work of Lesley Blankenship-Williams (author) unless otherwise stated below. For images used more than once, only the first occurrence is listed here.

Front Cover image: Adobe Stock Images

Page 5: Photo taken by Palomar College TV Center.

Page 22: Water molecule graphic released into public domain through Wikimedia commons. Author unknown.

Page 28: Lung image by *Rastrojo* and released into Wikimedia Commons for use with attribution.

Page 29: Red blood cell image by Drs. Noguchi, Rodgers, and Schechter of NIDDK through NIH (a governmental agency) and therefore part of public domain.

Page 30: Coffee image by Petr Kratochvil and released into public domain. Water drop before impact photo taken by Roger McLassus in 1951 and released into Wikimedia Commons for use with attribution.

Page 42: Mr. Potato Head at Disneyland photo released into Wikimedia Commons for use with attribution. Author goes by *Freddo*.

Page 51: Black and white drawing of inner ear released by NIH into public domain. Colored graphic of inner ear created by Blausen.com staff and released into Wikimedia Commons for use with attribution.

Pages 52 - 61: All illustrations by Nalini Biggs. Illustrations were commissioned specifically for this book. © All rights reserved.

Page 64 & 65: Glucose, fructose and sucrose chemical structures released into public domain through Wikimedia Commons. Author goes by *Neurotiker*.

Page 66: Glycogen structure from the following article and released into public domain: E. Meléndez-Hevia, R. Meléndez and E. I. Canela (2000) "Glycogen Structure: an Evolutionary View", pp. 319–326 in Technological and Medical Implications of Metabolic Control Analysis (ed. A. Cornish-Bowden and M. L. Cárdenas), Kluwer Academic Publishers, Dordrecht. Kids holding hand artwork released into public domain.

Pages 69 & 70: Nucleotides modified by author from Cytosine and ATP images released into public domain.

Page 73: Image taken from https://www.idtdna.com/pages/decoded/decoded-articles/core-concepts/decoded/2011/03/16/unraveling-rna-the-importance-of-a-2'-hydroxyl and used with permission.

Page 74: "DNA Macrostructure" Illustration from Anatomy & Physiology, Connexions Web site. http://cnx.org/content/col11496/1.6/, Jun 19, 2013. Released into Wikimedia Commons for use with attribution.

Page 77: *E. coli* image uploaded by Magnus Manske and released into public domain. HIV image by J. Robert Trujillo and released into Wikimedia Commons for use with attribution.

Page 78 & 79: Amino acid images shown in Problem 2 and Problem 4 released into Wikimedia Commons for use with attribution by user *Borb*.

Page 81: ATP Synthase (Figure 3A) uploaded into public domain. Image from "1Q01" in www.rcsb.org. Original structure from following citation: Molecular architecture of the rotary motor in ATP synthase. Stock, D., Leslie, A.G.W., Walker, J.E. (1999) Science **286**: 1700. Cartoon drawing of ATP Synthase (Figure 3B) uploaded into public domain by user Matthias M.

Page 83: Upper image is AlphaHelixProtein.jpg released into public domain by NIH. Lower image is credited to Paul RHJ Timmers and released into Wikimedia Commons for use with attribution.

Page 86: Image released into public domain by NIH. http://www.nhlbi.nih.gov/health/health-topics/topics/sca/

Page 92: Illustration by Nalini Biggs specifically commissioned for this book. © All rights reserved.

Page 94: Mako shark jaw photo attributed to "Isurus oxyrinchus Machoire" by Didier Descouens. Port Jackson shark jaw photo attributed to "Port Jackson Shark Jaw 1" by user Jason7825. Both images released to Wikimedia Commons for use with attribution.

Page 95: "Erythrocyte oxy" image released into Wikimedia Commons for use with attribution by author *Rogeriopfm*. "Macrophage in the alveolus" image released into public domain by Louisa Howard and further modified by the author.

Page 96: Top illustration by Nalini Biggs specifically commissioned for this book. © All rights reserved.

Page 109: "Androstenedione" and "Cholesterol" images released into public domain by *Lukas Mizoch* and *Neurotiker* respectively: adreostenedoine image modified by author.

Page 111: Both illustrations from Anatomy & Physiology, Connexions Web site. http://cnx.org/content/col11496/1.6/ by *OpenStax College* released into Wikimedia Commons for use with attribution.

Page 114: Plasma membrane images released into public domain by Mariana Ruiz (aka "Lady of Hats").

Page 127: "Iron bacteria in surface water" photo released into public domain by www.nhep.unh.edu.

Page 129: Cartoon generated by Nalini Biggs specifically commissioned for this book. © All rights reserved.

Page 132: Car image by author *Vander01* was released into Wikimedia Commons for use with attribution. Modifications to image by author.

Page 134: Illustrations "Electron Transport Chain" from Anatomy & Physiology, Connexions Web site. http://cnx.org/content/col11496/1.6/ by *OpenStax College* released into Wikimedia Commons for use with attribution.

Page 135: Mitochondria image released into public domain by Mariana Ruiz (aka "Lady of Hats").

Page 144: Endocytosis image released into public domain by Mariana Ruiz (aka "Lady of Hats"). Bubble blowing photo "The Time Pass" by Sagar Josh is released into Wikimedia Commons for use with attribution.

Page 148: Image is "Osmose en" by Hans Hillewaert released into Wikimedia Commons for use with attribution. Image modified by author.

Page 155: Photo "Young patient with IV drip" by Mike Blyth released into Wikimedia Commons for use with attribution.

Page 156: Image by Blausen.com staff. "Blausen gallery 2014". *Wikiversity Journal of Medicine*. DOI:10.15347/wjm/2014.010. ISSN 20018762 released to Wikimedia Commons for use with attribution.

Page 161. Photo (http://www.picdrome.com/albums/food-and-drinks/unrefined-sugar.jpg) is public domain and further modified by author.

Page 163. Figure 2A, B. Enzyme images created from www.rcsb.org protein identifier 1LYD. Figure 2C Image from the RCSB PDB September 2000 Molecule of the Month feature by David Goodsell (doi: 10.2210/rcsb_pdb/mom_2000_9) and reprinted with permission.

Page 164. Pepsin from the RCSB PDB December 2000 Molecule of the Month feature by David S. Goodsell (doi: 10.2210/rcsb_pdb/mom_2000_12). Figure 3 is a collage of four images: DNA Helicase is from the RCSB PDB December 2013 Molecule of the Month feature by David S. Goodsell (doi: 10.2210/rcsb_pdb/mom_2013_12); RNA Polymerase is from the RCSD PDB April 2003 Molecule of the Month feature by David S. Goodsell (doi: 10.2210/rcsb_pdb/mom_2003_4); Topoisomerase is from the RCSD PDB January 2006 Molecule of the Month feature by David S. Goodsell (doi: 10.2210/rcsb_pdb/mom_2006_1); DNA Methyltransferase is from the RCSD PDB July 2011 Molecule of the Month feature by David S. Goodsell (doi: 10.2210/rcsb_pdb/mom_2011_7)

Page 165: Image by Thomas Shafee and released to Wikimedia Commons for use with attribution. Chemical reaction drawing released into the public domain by *Jmun7616*.

Page 166: Taffy photo by *Lara604* and released to Wikimedia Commons for use with attribution.

Page 173: Enzyme image created from www.rcsb.org protein identifier 1HW6 and modified by author.

Page 176: Nitrogenase from the RCSB PDB February 2002 Molecule of the Month feature by David S. Goodsell (doi: 10.2210/rcsb_pdb/mom_2002_2)

Made in the USA
Columbia, SC
26 July 2019